Near the Forest, By the Lake

Near the Forest, By the Lake

Discovering Nature Close to Home

Angela E. Douglas

Comstock Publishing Associates
an imprint of
Cornell University Press

Ithaca and London

First published 2025 by Cornell University Press

Library of Congress Cataloging-in-Publication Data

Names: Douglas, A. E. (Angela Elizabeth), 1956– author.
Title: Near the forest, by the lake : discovering nature close to home / Angela E. Douglas.
Description: Ithaca : Comstock Publishing Associates, an imprint of Cornell University Press, 2025. | Includes bibliographical references.
Identifiers: LCCN 2024023151 (print) | LCCN 2024023152 (ebook) | ISBN 9781501780370 (paperback) | ISBN 9781501780561 (pdf) | ISBN 9781501780578 (epub)
Subjects: LCSH: Natural history—New York (State)—Ithaca Region—Popular works. | LCGFT: Essays.
Classification: LCC QH81 .D676 2025 (print) | LCC QH81 (ebook) | DDC 508.747/71—dc23/eng/20241127
LC record available at https://lccn.loc.gov/2024023151
LC ebook record available at https://lccn.loc.gov/2024023152

Contents

Preface

I walked into the hotel room assigned to me as a member of the review panel. Instantly, I realized that something was awry. The room was in good order—the usual desk, the bed with too many pillows, and the cabinet bearing an enormous TV and a flimsy coffeemaker in matte black to conceal the coffee stains. Well-practiced, I first switched off the TV that was advertising special deals for sister hotels in exotic places and then shut down the AC unit that was blasting dusty Arctic air. The window was hermetically sealed as always, but the view of a damp Washington, DC, street was reassuring. It was different from the view of the parking lot I had from the previous week's conference hotel in San Francisco. Apart from that, everything was identical in the two hotel rooms. Even the abstract picture hanging on the wall opposite the window was the same confusion of blobby gray, green, and yellow vertical lines.

I felt weighed down; this was not where I wanted to be. Slowly, I unpacked my travel bag and tended a few urgent messages in my email inbox. I walked back to the window. I noticed a patch of green bushes on the other side of the road, beyond the intersection. Was it the edge of a city park? I reached for my coat. It would be chilly in the late-afternoon drizzle.

By the time I returned through the automatic glass doors of the hotel, I was back to my normal self. I felt positive and motivated by the work tasks ahead. I noticed that the hotel was clean and quiet and that the staff were doing their best. Once again, making contact with the natural world—in this case, a nondescript city park that majored in worn grass, paved paths, trash cans, and a children's playground—had made all the difference.

I find myself reflecting on this incident many times, perhaps because the identikit hotel rooms on opposite sides of the continent were a bit creepy, or perhaps because I would never have expected the small park to have the power to lighten my mood so dramatically.

My meandering thoughts that start with this Washington, DC, experience always end up in the same place: What do I truly value about the natural world? To my consternation, my attempts at a succinct answer to this question are invariably unsatisfying. The real answer, at least for now, is the set of essays in this book, which were written between summer 2021 and early 2023. It is a highly personal answer that is particular to the beautiful world of upstate New York, centered on the city of Ithaca, where I live. In this world, the local city park borders the southern end of Cayuga Lake, and it has osprey nests, bluebirds, unmown meadows of summer flowers, and butterflies. Yes, I am very fortunate.

The other and perhaps greater purpose of these essays is to entertain and remind ourselves that, despite the many insults inflicted by humans, the natural world is the most interesting and fun place to be.

Acknowledgments

I thank Ben Johnson for his thoughtful comments and suggestions for the essays on mole salamanders (March,) and spring peepers (April); Xiangtao Xu for valuable advice on carbon budgets of forested landscapes (October); and Paul Curtis for helpful advice on black bears (January) and for agreeing to the inclusion of the quotation from his email message. The Belle Sherman Writers Group (Jeff Barken, Val Bunce, Ann Gold, and David Levitsky) provided useful advice on a draft of the introduction, and John Ewer's comments relating to the postscript were very valuable. I am grateful to two anonymous referees, whose excellent advice has improved the manuscript, to Kitty Liu, the editorial director of Comstock Publishing, for her invaluable advice and support, and to Karen Hwa and the production team at Cornell University Press for their superb work. Above all, I thank my husband, Jeremy Searle, the partner in the "we" of my essays, for his boundless enthusiasm for the local natural world and his unfailing encouragement and thoughtful comments on my writing.

A modified version of the essay "Lilies in February" (February) was published in the online magazine *Monologging* (https://mono logging.org/say-it-with-flowers2), without any copyright restrictions for republication.

Near the Forest, By the Lake

Introduction

> It is interesting to contemplate an entangled bank,
> clothed with many plants of many kinds, with birds
> singing in the bushes, with various insects flitting
> about, and with worms crawling through the damp
> earth.
>
> *Charles Darwin,* On the Origin of Species

D arwin's dreamy words speak to the immense power of the natural world to please us. What did the author have in mind? I imagine a long, grassy slope near his home in the south of England. It is early summer, and the cowslips are in flower, a glorious lemon yellow, along with some purple dog violets. You can hear the buzz of bumblebees and the scratchy song of a white-throat singing from a nearby hawthorn bush, which is ablaze with clusters of pure white flowers. Swallows are flying low across the bank, and a song thrush takes off, its beak full of worms for its nestlings.

It is easy to change the scene at this entangled bank. Take it to a different season, perhaps to August when armies of small bees are visiting the delicate, pale blue harebells, and meadow brown butterflies are sipping nectar from the pink brushlike flowers of thistles. Change the players again by traveling to a grassy slope on a different continent. Perhaps our entangled bank can shape-shift altogether to a sphagnum bog in a conifer forest or the Arctic tundra; to an exposed rocky seashore where limpets and barnacles flourish, despite the onslaught of battering waves; to a coral reef resplendent with fish of every color; or to the cold, dark world of the deep sea where lanternfish make their own light of yellow, green, or blue, and giant squid capture their prey (including lanternfish) with thirty-foot-long feeding tentacles. An entangled bank is anywhere that is teeming with life in all its splendid diversity. Even the totality of animals, plants, and other organisms that make up the biosphere of this planet can be considered as one great entangled bank.

I am confident that I am not extending the meaning of Darwin's words beyond his intention. Charles Darwin did not invite readers to imagine an entangled bank as a rhetorical flourish about the diversity and beauty of the natural world in his local area. The epigraph at the beginning of this introduction is the opening of the final paragraph of *On the Origin of Species* (Darwin [1859] 1985), one of the most consequential books ever written.

For better or worse, the fame of *On the Origin of Species* does not relate primarily to the diversity of life but to its central thesis that no organism is here by design. All the animals, plants, and microbes alive today are the descendants of organisms that happened to produce the most offspring, a process that Darwin called natural selection. This carefully argued conclusion is still disputed, despite the overwhelming evidence that Darwin's ideas are fundamentally correct. The enduring controversy is the probable reason why subsequent parts of his argument that concern the complexity of the natural world tend to be neglected.

In principle, Darwinian natural selection could be a recipe for "winner takes all." The creature that produces the most offspring will eliminate the entire opposition and become the sole surviving species, a so-called Darwinian Demon. This line of reasoning is much beloved by certain politicians and some traditional economists, but it does not apply to the natural world. The difficulty for any budding Darwinian Demon is that there is never quite enough of what an organism needs (e.g., food) and too much of what it could do without (e.g., extreme temperatures, predators, deadly pathogens). What's more, many organisms have come to depend on partnerships with other organisms. For example, insects pollinate flowering plants in exchange for sugar-rich nectar, and pilot fish promote the health of sharks when they feed on the parasites infesting the sharks' skin. As Darwin recognized, the struggle for survival generates complex networks of interacting organisms or, in his words, the players on his entangled bank.

There is one more step in Darwin's long argument. Namely, the participants and interactions in the entangled bank are endlessly changing both in ecological time because of shifts in the conditions (e.g., weather, seasons) and interactions with other organisms (e.g., prey, pathogens, mutualists) and in evolutionary time because, down the generations, organisms with slightly different inherited traits produce more or fewer offspring. The consequence is elegantly summed up in the closing words of *On the Origin of Species*: ". . . endless forms most beautiful and most wonderful have been, and are being, evolved" (Darwin [1859] 1985, 460).

The words "most beautiful and most wonderful" reveal something more: Darwin's profound awe of the natural world. This emotional response is shared by many who spend time in nature (with cell phone off and earplugs removed). Some people experience heightened spiritual awareness or a sense of union with a greater power. The natural world can be a gateway to religious insight. In his poem "The Windhover," Gerard Manley Hopkins used the kestrel "rung upon the rein of a wimpling wing" as a symbol of Christ "a billion times told lovelier, more dangerous" (Hopkins 1888). Olivier Messiaen, who composed the monumental musical cycle of *Catalogue d'oiseaux* in 1956–58, combined an unswerving commitment to an authentic representation of the song of birds in their natural habitat with a conviction that birds are truly God's angels on earth. Darwin's awe of the natural world took him in a different direction. Darwin harnessed his deep knowledge of animals and plants to inform his scientific endeavors, revealing that the natural world can function without divine intervention.

A knowledge and appreciation of the natural world, as expressed so eloquently by Charles Darwin, are the central concerns of the discipline of natural history, the focus of the essays in this book.

Natural History

Natural history, sometimes referred to as nature study, is a descriptive activity that delights in the details. We observe creatures in their natural habitats, we may identify them, and we note details about their life cycle stage and, especially for animals, their behavior. Natural history is all about the particular and the personal. Put differently, natural history does not seek an understanding in terms of mechanism or process, which is the purview of science, an activity whose ultimate goals are not focused on either the particular or the personal. Nevertheless, natural history is closely related to science because observations about the natural world can play an important role in supporting or refuting scientific hypotheses. After all, Charles Darwin used his extensive knowledge of natural history to validate his scientific theory of evolution by natural selection.

Fortunately, natural history is an activity in which anyone can engage. Formal training is not needed, although we can all learn from an accomplished naturalist who, over the years, has become skilled in species identification and knowledgeable about the habits of animals and plants. By contrast, a scientist is a highly trained professional who has studied over many years. This difference between natural history and science may be blurred by the very welcome development of citizen science, enabling nonprofessionals to participate in the scientific process, especially with data collection and analysis. Nevertheless, citizen science programs are most productive when they are designed and run by professional scientists.

Charles Darwin was both a superb naturalist and a premier scientist. He was also very interested in humans, primarily from an evolutionary perspective. *The Descent of Man, and Selection in Relation to Sex* (Darwin [1871] 1981) provides thoughtful deliberations on the evolution of human anatomy and on the human mind and morality. (The book is, however, marred by references to race and gender in terms that are totally unacceptable today.) However, Darwin did not give any extended consideration to the place of humans in his entangled bank.

How times have changed! Today, humans figure prominently in any discussion of the natural world, including from the perspective of natural history.

Humans and the Natural World

There is a version of human life that is disengaged from the natural world. The habitat is made of climate-controlled buildings of perfectly vertical and horizontal surfaces and ninety-degree angles, connected by fast-moving metal boxes, such as cars and airplanes. To meet material needs, products are selected from images on a computer screen and delivered to the door. In this dystopian but not necessarily unfamiliar existence, we can be totally ignorant of where our food, clothing, and possessions come from and how they were made. A moment's thought reveals that such disengagement is an illusion. The raw materials for everything that humans consume or manufacture come from other living organisms and inanimate resources, including water, minerals, and the atmosphere. This is basic economics. It is a mystery to me why some professional economists treat our planetary resources as infinite and, consequently, worthless.

There is more to the human relationship with the natural world than economic transactions. We also benefit from direct contact with nature. The intuition that the natural world can make us feel healthier and happier is backed up by hard data obtained by academic researchers in many disciplines, from biomedical science to psychology, epidemiology, and the social sciences. Even after confounding factors, such as access to clean air, the opportunity to exercise, and disparities in wealth, are accounted for, access to nature gives a small but statistically significant boost to our physical and mental health and to life expectancy (Hartig et al. 2014; Jimenez et al. 2021).

We have reached a contradiction. Humans have a deep affinity for the natural world, but, as societies and individuals, we tend to adopt lifestyles that distance ourselves from nature. A useful way to think about how this has happened is to consider our origins. The trouble probably started at the time of the earliest complex societies in which individuals specialized in different tasks and took up writing to supplement communication by the spoken word. People embedded in a complex, "civilized" society have weakened day-to-day contact with the natural world. Archaeological records place the earliest human civilizations at

six thousand years ago, meaning that humans have been semide-
tached from the natural world for no more than 2 to 6 percent
of the time since our species, *Homo sapiens*, first evolved (one
hundred to three hundred thousand years ago). Fast-forward
to today. More than half of the eight billion humans alive today
live in cities, and some cities are very large. At least five hundred
cities have more than one million inhabitants, and the population in nine cities exceeds twenty million inhabitants (Gillespie
et al. n.d.). Billions of people adapted to their ancestors' world of
nature are today living in urban habitats where nature is mostly
squeezed out.

Furthermore, humans are not passive recipients of the goods
and services of nature on which we depend for our survival and
well-being. Rather, we have profoundly altered the natural world.
The scale of our interventions is so great that it can be difficult to
comprehend. Half of the world's habitable land has been requisi-
tioned by humans for agriculture (Ritchie 2024), and an additional
35 percent of the remaining land is not actively managed but has
been changed sufficiently to qualify as "novel ecosystems" (Marris 2011). The artifacts (e.g., buildings, bridges, dams) constructed
by humans weigh more than all living organisms on the planet,
including the world's forests, the entire human population, and all
other animals (Elhacham et al. 2020).

Many of the anthropogenic changes (changes caused by human
activities) can be attributed to self-interest. Forests are felled for
timber and to create agricultural land; waterways are dammed to
enhance human water supplies and for hydroelectric energy; and
animals are hunted and plants collected for food and other valued products (e.g., traditional medicines, furs, feathers), bringing
many species to the brink of extinction—and sending some over
the brink.

Other anthropogenic changes are incidental. "Exhibit A" must be our vast global experiment in anthropogenic climate change caused primarily by burning fossil fuels. For a century, the consequences of this activity were not appreciated. However, over the last thirty years, scientists have been issuing increasingly urgent warnings about increases in global temperatures and regional changes in climate caused by the uncontrolled emission of greenhouse gases (United Nations 2023a). The responses to these warnings are generally inadequate. For example, an analysis issued in late 2023 concluded that the goal of restricting the global temperature increase to less than 1.5°C above preindustrial levels could not be met by the climate action plans of most nations (United Nations 2023b). At a societal level, we cannot say that we don't know what we are doing.

The ramifications of climate change for the world's fauna and flora are incalculable. The winners will be those that can tolerate shifts in conditions (e.g., temperature, precipitation) or can disperse to different locations with more suitable conditions. The players in all the interconnected tangled banks of the world are changing, and they increasingly include novel combinations of species, often with unpredictable ecological consequences.

A different motivation for human modification of the natural world has become increasingly prominent in recent decades: to protect habitats and species that we value. We recognize the ecological importance, for example, of the forests and wetlands that serve as carbon sinks and of the insects that pollinate our crops; environmental economists refer to these activities as ecosystem services, giving them monetary value. We perceive certain landscapes, such as high mountains and forest-lined lakes, as beautiful and awe-inspiring, and we know that many species are endangered. A few are magnificent, such as tigers, or are endearing, such as pandas, but don't we all have a sense of loss, even if only fleetingly felt, when unremarkable species, such as a small frog or straggly shrub, goes extinct as a direct result of human greed and thoughtlessness?

Self-interest, ignorance, or a desire to make good the previous environmental insults: whatever the reason, humans are mighty meddlers of the natural world.

Humans as Mighty Meddlers

Humans are not the only meddlers of the natural world. American beavers fell trees, which they use to dam a stream, creating a pond. This suits the beavers, which often construct their dens with sticks and mud in the pool, well-protected from predators, and store branches underwater in the pool, as food through the winter. Inadvertently, beavers can drastically alter their environment by slowing water flow, raising the water table, and creating wetlands. Other ecosystem engineers are biochemists. The cyanobacteria are an ancient group of microbes that hit upon the chemical trick of photosynthesis to make sugars with oxygen as the waste product. This trait evolved about 1.6–1.9 billion years ago and, slowly but surely, the waste oxygen accumulated, profoundly changing the atmosphere of the planet. Chemical engineering of the planet's ecology by cyanobacteria created the opportunity for the evolution of organisms that can use oxygen as a fuel in respiration. Oxygen-respiring organisms, including humans, now dominate the planet. Beavers, cyanobacteria, and other meddlers are known as ecosystem engineers (Jones et al. 1994).

Nevertheless, human meddling is different from other ecosystem engineers in two respects. First, we meddle in many different ways. For example, we alter water courses (like a beaver), make a myriad of other structural changes to the environment, and modify the chemistry of the planet by pumping carbon dioxide and other greenhouse gases into the atmosphere, by interfering in the global nitrogen cycle, and various other interventions. Second, we have the capacity to reason. A beaver cannot reflect on whether it is good for the ecosystem to build a dam in a particular location, or at all, but we can weigh the economic, environmental, and public health effects of building a highway through a pristine habitat or polluting the land and waters with toxic chemicals and plastic waste. We can also debate the pros and cons of different strategies to reduce the ecological footprint of cities and agriculture, to restore a degraded habitat, or to protect an endangered species.

Decision-making about environmental issues is complex and often difficult. The challenges are not limited to habit and the vested interests that tend to favor the status quo. Even the best decisions founded on scientific information involve value judgments that balance financial feasibility, public health priorities, aesthetics, ethics, and what members of society value about the natural world. These considerations raise the question: How can we, as individuals, identify what is important to us in the natural world?

This brings me to the purpose of the essays in this book: to celebrate our engagement with the natural world through the discipline of natural history not only because this activity is fun but also because natural history helps us understand what we value about the natural world.

To summarize, science gives us concepts and explanations, and natural history—which, as explained above, is particular and personal—reveals the value of the natural world to us as individuals.

A Map and Compass for the Essays

This collection of forty-eight natural history essays focuses on the natural world in my local area around the city of Ithaca in upstate New York. In my explorations of local nature, I am usually accompanied by my husband, Jeremy Searle—hence the "we" in many of the essays.

Three themes permeate the essays, but first, I want to alert readers to a theme that is conspicuously absent: a temporal baseline. As Darwin appreciated, the natural world is always changing, and it is doing so now at an unprecedented rate because of anthropogenic climate change and environmental degradation. Many people see the natural world through the lens of alterations since their childhood, which is their baseline for what they perceive as natural. I lack this perspective because I spent the first half century of my life in Britain, meaning that I tend to compare between "here and there," not "now and then." I witness the natural world in my new home without any personal remembrance of times past.

Now for the three recurring themes, which some readers may find useful as a map and compass for the essays.

I value the predictability of seasons and the chaotic fluctuations in weather. This is true most of the time, although I admit to doubts when I am scraping ice off the car windshield on a bitterly cold February morning or when I am drowning in sweat after a five-minute walk up the road at midday in early August. The joy of a world dominated by season and unpredictable weather is that life does not stand still, and it is impossible to get bored. This month is different from last month; at some times of year, the natural world has noticeably changed since last week, occasionally since yesterday or even this morning. Nevertheless, the perpetual flux has a reassuring predictability. What has gone will come around again.

You could say that time is an arrow that, every year, bends into a circle. Reflecting this value, my essays are arranged by month, from January to December.

I value the human factor. From the time when I arrived in Ithaca in 2008, I knew that I valued the role of ice and water in creating the landscapes that dominate the local area: the glaciers that gouged out riverbeds to form Cayuga Lake and the other Finger Lakes and the creeks that eroded soft sedimentary rocks to create the gorges and waterfalls. It took me longer to understand and appreciate the human meddling. The landscape has been modified for millennia by Indigenous people who used fire extensively to optimize the environment for their lifestyle as hunter-gatherers with shifting agriculture (Jordan 2022). The settlement of the local area by people of European descent in the late 1700s was followed by large-scale forest clearance for timber and agriculture, and most of the forest today is secondary growth on agricultural land abandoned in the nineteenth and twentieth centuries.

Humans are also responsible for alien plants, mostly of European and Asian provenance, in the local flora. Many of the plants brought to North America, either by design, for culinary or medicinal use or as garden ornamentals, or by accident, for example as contaminants of imported seeds, have thrived, self-propagating in the wild. Laws that regulate the import and purchase of certain invasive plants and the welcome recommendations that gardeners "plant native" are reducing the influx, but various aliens have become an integral part of our local world. Clumps of Japanese day lilies in woodland areas and swathes of West European purple loosestrife in wetlands are common summertime sights. As I have already considered, further change is inevitable, particularly in response to anthropogenic perturbations to the climate.

The bottom line is that human activities, past and present, play a large role in shaping the natural world we enjoy and value.

I value the diversity of the natural world. Local nature is immensely varied. It includes a diversity of habitats, from hemlock groves and maple-beech woodlands to sedge-grass wetlands, grassy meadows, creeks, beaver ponds, lakes, and deep-plunge pools at the base of waterfalls. These habitats support diverse communities of animals and plants, including many species that are seasonal. Our local bird fauna is composed of both year-round residents and many winter visitors, summer visitors, and spring/fall migrants passing through. Some mammals and most amphibians and insects "disappear" into dormancy during the winter. Plant and butterfly species vary from early spring through high summer to late fall. What's more, the total diversity of my local natural world is a tiny portion of the great diversity of nature on this planet.

These three themes bind my essays and have given me insight into what I truly value about nature. The myriad of entangled banks and the endless forms of the natural world are truly awesome and have an immense power to please.

A final map-and-compass issue is that the essays can be read in any order. There is no linear plot that requires the earlier months of the year to be read before the later, and I cross-reference between linked topics covered in different essays. I conclude with a brief postscript that revisits the central theme of this book, the importance of the natural world in our lives.

January

The year starts in the heart of winter, and it is a spectacular time. At its best, the world of January is cold and crisp, blanketed in a layer of snow that sparkles like diamonds in the sunshine and gently muffles the sounds of everyday life. Considered differently, January, being the coldest and snowiest month on average, is not a time for wimps. Survival requires a tolerance of both temperatures below freezing for days, even weeks, at a time and regular periods of inactivity enforced by blizzards, freezing rain, high winds, and so on. However, nothing is certain; in some years, we have brief respites of unseasonably warm weather.

Much of winter-resident life hides away, dormant or in hibernation. Inescapably, the creatures that remain active, especially the birds, play a large role in our January world—and in my essays for this month. The year starts with a treat of turkeys and bluebirds on a trip to a nearby preserve ("New Year's Day Birds") and ends with flocks of overwintering duck on Cayuga Lake ("Luxury Living on the Lake"). In one year, January also brought a hawk into our neighborhood, causing trouble for the small birds at our bird feeder ("With Fear and Trembling"). Much more unexpectedly, a hike in the local forest revealed evidence for active black bears that should be hidden away in their dens at this time of year ("In the Company of Bears").

New Year's Day Birds

How do you mark the first day of the year? I gave up on New Year's resolutions years ago. Instead, I have developed the habit of saluting the new year by paying special attention to wildlife for the day. One New Year's Day was particularly memorable for the wild birds.

Not that the start was auspicious. We had a lonely breakfast. Our bird feeder, in full view from our dining room table, was deserted. No chickadees or tufted titmice, no juncos or Carolina wrens, not even the usual parade of marauding gray squirrels testing the squirrel-proofing. Perhaps the neighbor's cat was prowling about or a Cooper's hawk was watching from a nearby tree.

We decided to visit the Lindsay-Parsons Biodiversity Preserve, a dozen miles south of Ithaca. Our nil return for birds continued. It was already two hours after daybreak, and the thousands of crows that roost overnight through the winter in the trees on the southern outskirts of the city had already left for their daytime feeding grounds. Even the usual gaggles of ring-billed gulls that haunt the parking lots of the shops and eateries in the strip mall were not evident. It wasn't until we had traveled several miles south that we encountered our first bird of the year.

Driving around a corner, I had to brake quickly to avoid colliding with an unusual pair of pedestrians. Two wild turkeys were walking across the road in their own time, and they weren't going to be rushed. This meant that we had a splendid view. There's no argument: turkeys are seriously big birds; the great bulk of their body is supported by strong, powerful legs, and a ridiculously small head is perched atop a thin neck. The two birds were very similar in appearance, mostly a dullish dark brown with a pale face and thin, down-turned beak. As the birds moved, different areas of the plumage on their backs and wings glittered briefly in the hazy sunshine—the bird books call that an iridescent sheen. The turkeys dipped between the trees on the far side of the road. In a moment, they had disappeared, reminding us of how superbly camouflaged these great birds can be in their natural habitat.

Our two turkeys were probably females. If we had waited, we might have seen more turkeys, for these birds usually forage in fairly large groups during the winter. They like to scratch around for nuts, seeds, and berries on the forest floor, and they don't say no to any snails, beetles, or salamanders that their scrabbling may disturb.

It felt very special to have the wild turkey as our first bird of the year. We don't see turkeys very often, even though they are quite common in the country areas around here. It hasn't always been like this. The turkey was a victim of massive deforestation for timber and agriculture across the northeastern United States— and of uncontrolled hunting. By all reports, it had disappeared from New York State by the 1840s. But times change. Farming became unprofitable, and large areas of land were abandoned. As the scrub and trees grew up, wild turkeys returned. They expanded from small populations in northern Pennsylvania and, by 1950, became established in the wild areas of western New York State. Their subsequent spread was aided and abetted by staff from the New York State Department of Environmental Conservation who transported birds from their western stronghold, especially Allegany State Park, to other suitable parts of the state. Today, turkey populations are deemed sufficiently robust to support hunting throughout May and for two weeks in October, albeit with very strict bag limits per hunter.

The turkeys were the start of a great birdy morning. We were soon walking along one of the tracks close by the beaver pond at the Lindsay-Parsons Biodiversity Preserve. The trail was covered in snow and the pond was iced over, a reminder of the bitterly low temperatures and snowfall of the previous few days. One section was edged with brush and small trees. "What an excellent habitat for small birds," we said to each other. You bet! Nothing less than two bluebirds, a male and female, appeared as if from nowhere. I have read repeatedly that the plumage of the bluebird is "gorgeous," and that is the perfect word. The male was close by on a branch of a leafless sapling, his head and back the most gorgeous blue, and his throat and chest a glorious orange above a shockingly bright white belly. As he flew past us toward the female, the rich blue and orange flashed, as colorful as a butterfly. The female is more subdued, the blue-gray of the head merging to a stronger blue along the back and tail, and with an orange-brown chest. We are at the northern limit of the winter range for bluebirds. Perhaps the individuals we saw were resident to our area, possibly even breeding in the preserve, or perhaps they were winter visitors that will fly north in the spring to breed in southern Canada.

The bluebirds stole the show, but the regulars—including chickadees, tufted titmice, cardinals, and American goldfinches—were also busy in the trees and shrubs. At the far end of the preserve, American tree sparrows were twittering, a higher pitch and sweeter tone than the nearby goldfinches. It always feels special to come across the tree sparrows, for they are as tough as they come. They spend the summer months in the far north of Canada, migrating great distances to and from their overwintering grounds in the United States, as far south as Arizona and Texas.

Then a merlin flew through, fast and low. Merlins are small but immensely powerful predators, but they are not reported to over-winter in our region. Probably, this individual was passing through. Nevertheless, all the small birds (up to and including the cardinals) are in danger of becoming a meal for the duration of the hours, per-haps days, that the merlin was around. Only the crows and Canada geese feeding and calling on the open meadows had nothing to fear.

We weren't aiming for a record-breaking bird list for New Year's Day, but we did end up with a very reasonable tally. You can never be sure of what you will encounter in the natural world, but there is always something of interest, and it's often unexpected.

In the Company of Bears

Until this week, I could comfortably declare that I lived in a dif-ferent time and place from bears: both my old life in Britain, before I moved to the United States some fifteen years ago, and my new life in Ithaca, New York.

Let's start with "place." If our ancestors had as little wanderlust as our near-relatives, the gorilla and chimpanzee, we would still be in Africa, and much of the land mass in the northern hemisphere, including Britain and the United States, would be inhabited by large populations of bears. The most successful of these bears is the brown bear, *Ursus arctos*, which almost certainly evolved in Asia. It is a big beast, weighing up to 550 kg (1212 lb) and can wander far, up to fifty miles a day. So the brown bear has spread. It went across Eurasia in one direction, reaching Britain soon after the last ice age, and across Beringia (now the shallow sea between Siberia and Alaska) to North America in the other direction. With such a broad range, it is hardly surprising that the taxonomists have been busy dividing the brown bear into subspecies of greater or lesser validity. The brown bear in North America is *Ursus arctos horribilis*, the grizzly, and the Eurasian brown bear is *Ursus arctos arctos*.

You can tell from this that brown bears aren't especially fussy about where they live. They get along nicely in any type of forest or partially open habitat, from the frozen expanse of northern Russia to North Africa, from chilly Alaska to balmy Mexico. Well, that present tense is not strictly accurate because the last known Mexican grizzly bear (sometimes known as *Ursus arctos nelsoni*) was found, and shot dead, in 1976.

This brings me to the "time" of my opening sentence. If I had been born four thousand years ago, my Bronze Age life on the offshore island of Britain would undoubtedly have been nastier, more brutish, shorter . . . and shared with *Ursus arctos arctos*. But large numbers of humans and bears don't mix well. Slowly but inexorably, the brown bear populations were squeezed by an unhappy combination of habitat loss, as the natural wildwood habitat of Britain was felled, and relentless persecution. Every downed brown bear translated into a lot of meat and fur, and fewer sheep and goats ending up as bear food. People did not care that bears are omnivores that enjoy plant roots, berries, and insects as much as wild deer, domesticated sheep, and leaping salmon.

I was born too late to share my UK life with bears.

What about my US life? I would never have shared my life in the northeastern United States with a brown bear. The brown bear, or more specifically the grizzly subspecies of the brown bear, ranged widely across the west but never set up home east of the Mississippi River. In the last two centuries, the numbers and range of the grizzly have declined precipitously, eliminated from Texas (1890), California (1916), New Mexico (1931), and Colorado (1951). . . . I could go on with more state statistics, even providing you with the name of the "famous" person who made the last state kill, together with his location and details about his gun.

But the brown bear is not the only bear. The US is host to three of the eight bear species—the polar, black, and brown. It is time to bring the black bear—*Ursus americanus*—into the limelight.

The black bear is much smaller than the brown bear (300 kg = 660 lb at most). It is restricted to North America, and all it asks for is a varied menu of roots, fleshy leaves, berries, insects, and small mammals. Without interfering humans, it thrived in every scrap of forest on the continent. Alas, the modern distribution of the black bear is a pale shadow of its former range. It is now restricted to forests throughout Canada, along the main mountain chains of the US, and in a motley collection of other remote spots, such as the impenetrable swamps of Florida, Alabama, and Louisiana.

I moved to the US in 2008, specifically to an Ithaca that was the wrong time and place for black bears. Black bears had been extirpated from the Ithaca area more than one hundred years ago, and New York State populations were restricted largely to the Adirondacks in the northeast, the Allegany Mountains of the southwest, and the Catskills in the south. But, as I have already said, times change. The abandoned farms of the Finger Lakes region are being replaced by forest that includes, or is adjacent to, housing with garbage, pet food, and bird feeders—and the black bears are responding to the new opportunities. For at least twenty years, bear-free Ithaca has been wedged between a northern population expanding southward and a southern population expanding northward. In military terminology, this is the classic pincer formation, from which the sitting duck (the enemy, or the city of Ithaca) can be targeted from two sides simultaneously. There have been occasional reports of black bears seen here or there in our local area, but I've never been quite sure what to make of these stories.

So let's go back to where we began. My opening sentence starts with "until this week." It was this week that we went for a walk in one of our favorite haunts, Sweedler Nature Preserve, south of Ithaca. About ten minutes along the trail, I noticed a pile of wood fragments around the base of a dead tree, and portions of the trunk were torn away. If anything can be labeled as a frenzied attack, this was it! Other dead trees nearby had been attacked in a similar way, high up, low down, or scored from top to bottom. We puzzled over the likely perpetrator. A quick email consultation with Paul Curtis, an expert in the Department of Natural Resources and the Environment at Cornell University, yielded this response: "It looks to me that the damage was caused by a black bear, probably trying to get at grubs in the dead wood. With the mild winter weather so far, some juvenile male bears may not be in their winter dens yet."

This is why I wrote that, until this week, I could comfortably declare that I lived in a different time and place from bears. I have evidence that my world is different now; it has suddenly become wilder. I am living with bears.

With Fear and Trembling

We have a panoramic view of our small backyard from the dining room. Over breakfast yesterday, we admired the crisp and even snowscape glistening in the early morning sunshine, and we hoped that the temperatures would rise from the bitter $-9°F$ $(-23°C)$. The little birds must have burned through the calories to survive the previous night. Perhaps some of them went into torpor, allowing their body temperature to drop by a few degrees, and others may have huddled together in a tree hole for warmth.

Nevertheless, these clever tricks don't eliminate their intense demand for food at daybreak. Our winter birds don't indulge in fat stores to endure hard times; they need a constant supply of food. You could say that they are the true professionals of the just-in-time food supply chain.

If this were any previous year, we would have enjoyed a constant to-ing and fro-ing of small birds at our feeders as we ate our breakfast. Parties of chickadees and tufted titmice flying over the fence to take a few mouthfuls and then passing on to their next food stop; a Carolina wren craning its neck to access the food; a downy woodpecker bobbing along the branch toward the suet; a junco settled next to the bottom part of the seed holder to take a morsel, consider the world, and repeat; our resident pair of cardinals flying in to sample the seeds spilled on the ground by the flurry of small birds above. Perhaps we would enjoy a special visit from a red-bellied woodpecker or even a small group of pine siskins.

But every year is different, and this year is especially so. The entertainment for our morning breakfast was nothing more than a solitary junco and some gray squirrels chasing through the trees at the far end of the yard.

A likely explanation for the lack of interest in our bird feeders is that a natural predator has taken up residence, and our local patch has become a danger zone for small birds. The chief suspect is a Cooper's hawk. We have always enjoyed sightings of the occasional Cooper's hawk in our neighborhood, whether in swift and effortless flight across the yard or completely still, watching from the fence or a tree. This year, we have had one moseying around. We are sure it is the one bird; they tend to be solitary out of the breeding season, and they don't generally travel large distances in winter. It is almost certainly a juvenile because its orange-red chest is spotted, not striped, and the bands on its long tail are brown, not gray.

Our suspicions felt vindicated this morning. At breakfast, we spotted the villain of the piece. The Cooper's hawk swept low directly in front of our dining room window, a flash of blue-gray with pale underparts, then up and over the privet hedge to our western neighbor's backyard.

Until this winter, our birdy neighborhood watch has kept the hawks at bay. No self-respecting hawk would be willing to endure repeated mobbing by squawking blue jays and cawing crows. Somehow, the neighborhood protectors haven't been up to scratch this year. The consequence for our little birds is that a quick snack of Ithaca bird seed blend or a bite of suet has become an opportunity to dice with death. They are living in fear and trembling that they might end up as hawk food.

I can't help but wish our teenage thug of a Cooper's hawk would move on and stop tormenting our chickadees, cardinals, and other resident birds. In the same breath, I tell myself to be glad we have Cooper's hawks, along with sharp-shinned hawks and red-tailed hawks. It hasn't always been this way. From the early 1900s, these beautiful birds were persecuted by local farmers with increasingly accurate and deadly guns. The situation worsened in the 1950s with the widespread use of DDT, an organochlorine insecticide that disrupts the nervous system of insects. (For those who like the gory details, DDT is short for dichlorodiphenyltrichloroethane, and it stops the sodium channels in nerve cells from closing so nerves can't stop firing. DDT does this to the nerves of any animal, including you and me, but insects are unusually deficient in the enzymatic wherewithal to degrade DDT before it gets to their nervous system.)

DDT was the "magic bullet" insecticide of the 1950s, and its use led to great public health achievements. For example, it eliminated mosquito-vectored malaria from Mediterranean countries, such as

Italy, and southern US states, such as Louisiana and Alabama. Then evidence accumulated that the nervous system is not the sole target of DDT; reproductive hormones are also disrupted, albeit by mechanisms that are poorly understood. Alligators in Florida were found to have deformed testes and ovaries, and calcium deposition into the shell glands of birds was perturbed. The birds at the top of the food chain, the hawks and eagles, suffered the most. The thin shell of their eggs literally broke under the weight of the incubating parent.

It is thanks to the tireless work of many researchers and environmental activists that the disastrous effects of DDT became clear. Rachel Carson's *Silent Spring* (Carson 1962) was front and center in galvanizing proper controls over the use of chemical pesticides. Carson was a truly remarkable individual: a marine biologist who worked for some years at the U.S. Bureau of Fisheries and who clearly understood how the United States Department of Agriculture was hamstrung by its competing responsibilities to regulate pesticide use and promote agribusiness. Her work and the commitment of many others led to the banning of DDT in 1972 in the US. Since then, many birds of prey have made a remarkable recovery. The Cooper's hawk now has the conservation status of least concern.

So, as I observe the juvenile Cooper's hawk sitting perfectly motionless in the black walnut tree, watching and waiting for its next meal, I remind myself of the many people who cared enough to rescue its forebears from extinction.

Luxury Living on the Lake

There are various ways that you can live a luxurious life on the lake. I mean Cayuga Lake. Of all the Finger Lakes, Cayuga Lake is the longest, at nearly forty miles, and the second deepest, reaching more than four hundred feet in places. It is a little more than a mile wide and runs on an almost perfect north-to-south axis, with Ithaca at the southern end.

Please understand, by "luxurious life," I am not thinking of a luxury apartment or lake house with magnificent views over the lake. Rather, I am referring to living on the lake, getting-your-feet-wet style, as enjoyed by various birds that come from the north and stay with us for anything from a few days to the entire winter. Cayuga Lake in January may appear uninviting to us, what with the ice and the bitter wind, but, if you are born and bred in the far north of Canada, this is a balmy spot.

We have access to the shoreline at the southern end of Cayuga Lake, courtesy of two city parks: Stewart Park and Allan H. Treman State Marine Park. The birds on the lake are not unduly perturbed by the procession of dog walkers, joggers and, after snow, cross-country skiers. This means that amateur birdwatchers with binoculars, including us, and serious birders with telescopes join the merry throng of humans in these two city parks.

Recent days have been a birdy treat. A raft of ducks has taken up residence just offshore. Most of them are redheads, easily distinguished by the rich brown head and gray back of the males; the females are a dull brown. There are hundreds, maybe a thousand, redheads, all bobbing up and down on the water. Among the crowd of redheads, there are at least two other species of duck: some canvasbacks, which are a bit larger than redheads with white on the back, and scaups, which are smaller and black with white on the sides. (You would need a telescope to discriminate between the two species: the greater and lesser scaup.)

For many minutes, most of the ducks in the raft appeared to be doing nothing. Indeed, some had their heads turned back, as if snoozing gently. Occasionally, an individual rose and flapped its wings, or dived, presumably in search of food.

Then, all of a sudden, individuals at the far end of the raft started to get fidgety. They were swimming actively, initially in circles, then increasingly tending to face toward the center of the raft. A few of the uneasy birds rose up and moved closer to the center, powered by vigorous flapping of wings and furious paddling of feet. Their movement created streams of water droplets that glistened in the hazy sunshine. Soon, more ducks were on the move, more and more, until the far end of the raft was a mass of flapping wings in a shower of water. Some of the ducks left the water entirely. Most of these individuals flew fast and low for some tens of yards, then extended their legs diagonally forward as if to brace their body for the return to the water. A few rose into the sky, circled around, and returned to different locations on the edge of the raft. Their landings were splashy affairs, water skidding behind them as their forward-directed legs surfed along the water.

Then everything quietened down. We were back to normal, except that the raft had taken on a different shape. It was more rounded, perhaps more defensive. Throughout these maneuverings of the minority, most of the ducks in the center of the raft remained unperturbed, continuing to do nothing. That's luxury living on the lake.

The great number of redheads is awe-inspiring, but the top prize for elegance on the lake goes to another bird, the tundra swan. Its pure white body, long neck held strictly vertical, and jet-black bill were on full display as a small group of these magnificent birds glided over the water. Occasionally, an individual swan upended to take a mouthful of weed. The tundra swan also grazes on winter cereal crops in a similar way to the geese, but, so far this winter, we have not seen any swans among the large flocks of Canada geese feeding on the local snow-covered agricultural fields.

The chance to watch tundra swans feels very special because the main overwintering grounds for this species are on the coast and farther south, mostly along the Chesapeake Bay and in North Carolina. But tundra swans are strong flyers, and they move around a good deal through the winter. We have a chance of continuing to see them until mid- to late April. By early May, they need to be back to their breeding grounds on tundra lakes close to the northern coast of Canada.

A close second for elegance is a resident duck, the common merganser. In recent days, there have been dozens of them on the lake. The male has a dark green-black head, brilliant white neck and underparts, and gray-black on the wings, and the female has a chocolate brown head sweeping back into a punkish crest and gray-brown body. Most distinctive of all is their bright red beak, which is very long and thin, with a downward hook—a sure sign that the merganser likes its meat. Fish, worms, shrimp, and frogs go down easily. I suspect that the main item on today's menu is fish. That is because most of our local fish overwinter in the lake, a much safer place than the shallow, freeze-prone waters of the creeks . . . unless they meet a merganser, of course.

As we made our way through the snow at the lakeshore, we noticed the osprey platform silhouetted against the gray sky. A pair of osprey nest there every summer. It won't be so very long before the redheads and the tundra swan have departed north and the ospreys are back, patrolling Cayuga Lake and raising their young in this city park.

February

February spells "winter," just like January. The weather experts tell us that, in February, the average temperature is a little higher and snowfall is a little lower than in January, but the fluctuations around these mean values are so unpredictable that the average human detects no change. Much of the natural world also treats February as an extension of January. Lie low to survive the cold. But some creatures that are active through the winter are alert to increasing day length, the most reliable cue that spring is on its way. At our latitude, the time from sunrise to sunset is extended by a full hour, from ten hours to eleven hours, during the month. Every year, I am astonished that the dawn chorus of birds, hushed through the darkest days of winter, starts up in the wintry wastes of February.

With the February dawn chorus in mind, "The Sound of the Syrinx" celebrates the opening performer, the male cardinal, and the hardware that enables him and other birds to sing. This essay is preceded by an exploration of the remarkable strategies adopted by turtles and newts to survive the winter ("Living with Ice") and is followed by an examination of the bald eagles that overwinter on the lakeshore ("The Great Seal"). I conclude this chapter with some heartfelt praise of flowers ("Lilies in February"), which are sorely missed by late February.

Living with Ice

When the wind and snow barrel in from the high Arctic, our world becomes white and gray. The clouds occasionally part to reveal a sky of the lightest blue and a pale sun that makes the snow glisten. The ground is rock hard, the waterfalls are frosted into icicles, and the creeks and lakes are covered in a layer of ice that is the color of steel.

Ice. For us city types, ice is a peril of winter. When the air is well below freezing, a sheet of ice may be hiding under a cover of snow on the sidewalk. When the temperature hovers around the freezing point, both rain and recently thawed snow are transformed into an invisible but slippery glaze on the streets, sidewalks, and driveways. The only solution is salt, and lots of it, even though the underbelly of every car becomes rusted as a consequence. We are told to walk like a penguin to protect against bruised bottoms and broken limbs.

However, living with ice is not only about the rapid depreciation of the value of our cars and poor imitations of penguins. For many, living with ice means being trapped for weeks on end under a thick, icy layer. I have plenty of opportunities to think of the captives: when I stand on the bridge overlooking the pond at the Cornell Botanic Gardens, when I check the view of Cayuga Inlet (a creek that opens into Cayuga Lake) through the plate glass window of the indoor swimming pool, and when I walk the city park trail that runs alongside the frozen southern end of Cayuga Lake.

The ice captives include snapping turtles, which spend the summer swimming languidly through the murky waters of local ponds, and painted turtles, which bask in the summer sunshine on partially submerged logs. As the water cools in the fall, the turtles swim down to the muddy bottom, dig a hole, and bury themselves. The temperature down there drops to 39°F (4°C) for the entire winter. That's chilly, but it is a far better way to sit out the winter than on land, grappling with the hazard of getting so cold that ice crystals form in the blood or tissues, which would be curtains for a turtle.

What's more, a turtle doesn't have to worry about its energy bills because it turns its metabolism down to about 1 percent of the summer rate. This means it can manage without food for three to four months. Its oxygen needs are also trivial and can be met without breathing. It just hangs its head, legs and tail out in the sloppy mud, and lets the oxygen diffuse across its skin into its bloodstream. As the winter progresses, the oxygen supply in the mud gradually declines. No worries—both snapping turtles and painted turtles can switch to anaerobic respiration, without oxygen. You might imagine that this would lead to a buildup of lactic acid (like in your legs when anaerobic metabolism kicks in after running up several flights of stairs), which would acidify the blood and cause physiological mayhem. Our turtles deal with that complication by dissolving out some of the calcium from their shells, and the resulting calcium carbonate in the blood neutralizes the lactic acid. There is one last difficulty. It could get a bit boring, even claustrophobic, down there in the slime, but that is really no bother because the turtle's brain is operating in minimal maintenance mode, just like the rest of its body.

Another regular in our ponds are newts, specifically the eastern newt, also known as the red-spotted newt because of bright red spots encircled in black along the side of its otherwise dull green body. Unlike the turtles, newts remain active in the water column under the ice throughout the winter, although they are in the slow lane because of the cold. There's not a lot of food, but that doesn't appear to be an issue because newts can fall back on fat reserves laid down in the late summer and fall. No, the biggest concern for an overwintering newt is the oxygen supply under the ice because, unlike turtles, newts cannot switch to oxygen-independent anaerobic respiration. When a pond is covered in ice, the water is sealed off from the oxygen-rich air, and the only way to replenish oxygen is through photosynthesis by the aquatic algae and plants—but if the ice gets covered in snow, it can get very dark in the pond. When that happens, the oxygen gets depleted, resulting in what are known as newt winter kills.

It is certainly tough to be a newt. An added complication for the eastern newt is its unusual way of life. It starts off like other newts: as an egg attached to water plants in the spring, then as a tadpole with frilly gills and legs through the summer. The special quirk of the eastern newt is that, as the days shorten, the youngsters—known as efts—develop wanderlust. They crawl out of the water, resorb their gills, and, quite literally, wander the world for two or three years. These efts are bright red. We often see them crawling about in the forest during the summer. During the winter, they hibernate on land, sitting out the cold under logs and in leaf litter.

All this adds up to a realization that every adult newt we see in the early spring is a master of survival. It has survived at least two winters at risk of turning to ice in leaf litter plus at least one winter sealed under ice in danger of suffocation.

Our local turtles and newts put our human anxieties about ice into perspective.

The Sound of the Syrinx

During the second week of the month, the male cardinal in our backyard told the world that, for him, this was the first day of spring. It was dark and several degrees below freezing with a brisk northwesterly wind. About fifteen minutes before daybreak, he started calling and singing: *pip, pip, pip, chewee-chewee,* then a whistle ending in a fancy little trill. He kept it up until about ten minutes after dawn and then took a break, presumably for his breakfast. He has been back on singing duty every morning just before sunrise since then. Although both the male and female cardinals have been calling all winter, the predawn routine of the male is new. He must have decided that it was time to tell everyone and everything that this territory is accounted for and he survived the night. He did not fall prey to the local cat, the great horned owl, or the atrocious weather. Any other cardinals lurking around have received their daily marching orders: go and lurk somewhere else.

Most days, the cardinal has been our sole songster. Then the wind shifted briefly to the south and southwest earlier this week, bringing us a snatch of Florida weather. It wasn't record breaking, but it reached 60°F (15°C) with brilliant sunshine. Other birds also decided it was spring. The chickadees were calling, *chickadee-dee-dee* and *pheebee-pheebee,* along with the tufted titmice, *peeter-peeter-peeeterter.* Best of all, a mourning dove was perched high in the black walnut tree, his slim body, small head, and narrow tail silhouetted against the sky. I know it was a male because he was cooing, gently advertising his availability to any females in the neighborhood. As the sunshine warmed the air, the cooing became more insistent. Before long, there were male doves perched in conspicuous places up and down the street, all making the most of the springlike conditions. By this time, the doves didn't sound quite so gentle. Mourning doves are professionals in the art of competitive cooing.

And then there was another call: a gargling, then a swallowing sound, and a little whistle. I was mystified. I clutched at my phone for the Merlin Bird ID app to solve the problem. It was an eastern cowbird. Just as our songbirds and doves are getting into courtship mode, anticipating a happy family of nestlings, so the local cowbirds are setting up for another season of sneaking eggs into the nests of unsuspecting victims.

Of course, our experience of spring in February was brief. The temperatures have since plummeted, accompanied by snow showers, and we are back to the wintry wastes. The chickadees and tufted titmice twitter in subdued tones, the mourning doves are silent, and the cowbirds are gone. Nevertheless, our brave cardinal wakes us up daily as he sings in the darkness from the maple tree just outside our bedroom window.

From a musical perspective, our early season songsters are not so special. They are not nearly as loud as the Carolina wren, who will assuredly be singing before long, and their repertoire is not as melodious or complex as the catbirds and orioles that will be returning to us in early May. Overall, the total repertoire of calls and songs of most birds is surprisingly complex and loud, compared to mammals of similar size.

This is where the syrinx of my title comes in. Birds have abandoned the standard voice box of other land-living vertebrates: the mammals, reptiles, and amphibians. That is the larynx, which sits at the top of the windpipe and sports two little bundles of tissue called the vocal cords. For unknown reasons, the ancestors of the birds decided to delve deeper, all the way to the far end of the windpipe, where it divides into two pipes (bronchi) that take air into each of the two lungs. Birds don't depend on the vibrations of vocal cords to make a noise. Instead, they let the walls of the syrinx vibrate. Just as the pitch of the human voice is determined by muscles that shorten or lengthen the vocal cords, so the pitch of birdsong is dictated by muscles that alter the tension in the syrinx wall.

The important point about the syrinx is that it allows for bigger, better songs, but only because of a second anatomical innovation that birds are heavily (or should I say lightly?) into air sacs. A complex set of muscles alternately compresses and inflates the different air sacs, making them act like bellows, pushing and pulling air across the lungs. That is good for oxygen supply and the expensive business of flying, and it also creates a nice, steady airflow through the syrinx for making noises. What's more, these air sacs are great resonance chambers. The sound from the syrinx can be bounced off the walls of the air sacs, increasing the intensity of the sound, just like the sound board of a piano or the body of a violin.

There is more to a bird than its voice box. A bird must also breathe. Birds usually sing only when they are breathing out. Birds can sustain long songs, for example the extended trill of the cardinal, without keeling over from a lack of oxygen because they take mini breaths, up to twenty of them every second, that interrupt the sound for such short durations that it is not detectable. Of course, coordinating the muscles for breathing and singing is a very complicated business, especially since singing involves precise changes in the activity of multiple muscles that control the tension in the syrinx wall.

Let's add another layer of complexity. As I have already said, the syrinx of most birds is at the junction of the windpipe and bronchi.

This means that singing can involve parallel vibrations in the two bronchi. So why not make noises of different pitch in the two bronchi? No reason why not! Many birds do, making tunes with ascending and descending lines at the same time. The thrushes are incredibly good at doing this. You could say that a thrush is singing a duet with itself. But the syrinx is restricted to the bottom of the windpipe (it doesn't extend into the bronchi) in a few birds, including ovenbirds and brown creepers. Next time you see a brown creeper, remind yourself of something we have in common with them: we also can only vocalize just one pitch at a time.

The constraints imposed by having a syrinx in your windpipe are nothing compared to losing your voice box altogether. The turkey vultures that have recently returned from downstate and are soaring above our backyard are in this category. They are silent, apart from the occasional hiss and whine. Turkey vultures get along just fine without indulging in any cheeps, chirps, chatters, or croaks. The story of the syrinx reminds us that there's no single way to run a life, at least in the bird world.

The Great Seal

It has been an excellent winter for sightings of bald eagles. Several individuals have been hanging around at the southern end of Cayuga Lake for weeks. The birds that we have seen are immature, mostly brown with some dashes of white on the body. It takes four to five years for a bald eagle to acquire the adult plumage of a uniformly brown body with a brilliant white head and tail, plus a yellow beak and legs. Our overwintering birds could be born-and-bred locals or migrants from farther north.

It feels very special to be living alongside eagles. Somehow, the word *eagle* conjures up images of magnificence and power. The eagle symbolized Zeus, the chief god of ancient Greece, and it was the standard for every army legion of ancient Rome. Then it was adopted to represent St. John, one of the four Christian evangelists, and in the last five-to-six hundred years, the eagle has been used widely in heraldry, especially by monarchs in continental Europe. These symbolic eagles are all derived from the golden eagle, *Aquila chrysaetos*.

When the United States decided it needed a national symbol, it is hardly surprising that an eagle was chosen. What's more, one would expect the choice to be the golden eagle, which is distributed across both North America and Europe.

Despite this expectation, we all know that the eagle selected to represent the nation was not the golden eagle but the bald eagle. The story of how and why is rather convoluted.

We have to go back to July 4, 1776, and an assembly room in the Pennsylvania State House in Philadelphia where the Declaration of Independence was duly signed. What next? What is the first thing you would do after declaring unilateral independence from a superpower headed by a tyrant given to episodes of madness? I would call up any able-bodied person and get them armed in anticipation of an uninvited visitation by the trained army of the aforesaid tyrant. That was certainly on the to-do list, but our heroes thought bigger. Within hours of signing the Declaration of Independence, they set up a committee. Three key people were to spend their time designing a coat of arms, which would be the logo for the new country and the imprint for the wax seal appended to every important document. It would be the Great Seal.

The three talented men sent off to design the seal were none other than Benjamin Franklin, Thomas Jefferson, and John Adams. Unfortunately, none of them had attended the mandatory training course on teamwork run by the local personnel department.

They came up with three different ideas. Franklin went with the Egyptian king drowning with his chariot and horses in the Red Sea while Moses and the Israelites looked on from a high cliff; Jefferson and Adams preferred Hercules, although they differed on the details.

Unsurprisingly, Congress was unimpressed by the committee's offerings and set up a second committee. When that was not good enough, they set up a third committee . . . then a fourth. Exhausted by all of this, the fourth design, created by Charles Thomson and William Barton, was accepted, and the seal was in place in 1782. It took six years to design a logo. The central feature of the seal was a mostly brown bird that was suspended, as if on an invisible piece of string, with belly forward and wings outstretched. Its two legs were splayed out like arms, one clutching what looks like a bunch of flowers (but was actually an olive branch) with thirteen leaves and thirteen fruits, and the other holding a bunch of thirteen arrows. The bird's large beak grasped a yellow ribbon bearing the words *E Pluribus Unum* ("Out of Many, One"). Above this is a decorated circle with thirteen yellow stars. The nascent country was big on thirteen because, at that time, it was made up of thirteen states.

The bird on the Great Seal was an eagle—not a golden eagle but a bald eagle. The only clues to its identity were its white head and white tail. I suspect the designers had worked from a poorly prepared stuffed specimen.

Benjamin Franklin was incensed. He raged that the eagle looked like a turkey. In due course and after Franklin died in 1790, the US government responded to Franklin's complaints. The design of the seal was modified in 1825, then again in 1841 and 1877, but none of these quite cut the mustard. In 1885, yet another committee hit the sweet spot with a rather grand and formal bald eagle. The 1885 Great Seal remains unchanged to this day, adorning every $1 bill.

But we're getting ahead of ourselves. All that is still in the future. Let's get back to Benjamin Franklin, who was never shy about voicing his opinions. The poor artistry of the 1782 Great Seal was not his only problem. Franklin was dismayed that his country was adopting the bald eagle as its symbol of freedom and democracy. His disquiet is best expressed in his own words, as written to his daughter (Franklin 1784): "For my own part I wish the Bald Eagle had not been chosen the Representative of our Country. He is a bird of bad moral character. He does not get his living honestly. You may have seen him perched on some dead tree near the river, where, too lazy to fish for himself, he watches the labour of the fishing hawk; and when that diligent bird has at length taken a fish, and is bearing it to his nest for the support of his mate and young ones, the bald eagle pursues him and takes it from him. With all this injustice, he is never in good case, but like those among men who live by sharping and robbing he is generally poor and often very lousy."

Benjamin Franklin was an observant ornithologist. The bald eagle is principally a scavenger. Fish are the main item on its menu, and it favors dead fish over live fish. It is also very content to pirate food caught by other birds, including herons and osprey (that's Franklin's fishing hawk). Only when it is seriously hungry will a bald eagle deign to hunt for itself.

Unlike Benjamin Franklin, I don't see the bald eagle as a creature of bad moral character. Instead, I see a bird that has recovered from a close shave with extinction. Originally widespread in every state of the union, the bald eagle's habitat was reduced to a few remote locations around the Great Lakes and west of the Rocky Mountains by the 1970s. Its comeback over the last fifty years has been nothing short of astounding. In 2007, the bald eagle was removed from protection by the Endangered Species Act, and its population now exceeds 300,000 individuals.

Benjamin Franklin couldn't have known, but the bald eagle on the Great Seal has come to signify the astonishing capacity of nature to recover, if we only give it a chance.

Lilies in February

Now that the end of February is approaching, we are starting to tell ourselves that we are over the worst of the winter. The days are getting longer; the average temperatures are increasing, although still almost entirely below freezing; and . . . the self-talk fades away as we stare numbly at the winter scenery, the white of snow, and the black and gray of gaunt leafless trees silhouetted against the persistently dark gray skies in this flowerless land. That's the problem: it's flowerless! Lilies in February are as likely as a flying saucer landing in our backyard.

Experiments conducted by psychologists have demonstrated how flowerless is bad for us. Flowers make us happy; we smile and are more content in the presence of flowers. Flowers are even claimed to improve our memory and capacity to reason. What can we do in flowerless February? One solution is to use our imagination. We can think about flowers to keep up our spirits on the most miserable February day. In that spirit, here is a story about flowers.

Let's start at the beginning. That's about 170 million years ago at the height of the Age of Dinosaurs (also known as the Mesozoic Era), when something strange was happening in the green and pleasant land. The reason for all the buzz was a new kind of plant whose reproductive parts were decorated with modified leaves. It was the time of the first experiments with petals—and the first flowers. The fossil record tells us that flowers started off big, a bit like a water lily or a magnolia. The petals of the first flowers are usually portrayed as brilliant white, but I am unaware of any evidence that this is more than an educated guess.

The origin of flowers had an enormous effect on some of the insects that had previously made a living by eating the reproductive tissues, especially the abundant male spores, of flowerless plants. Many of these flowerless plants didn't just tolerate these insects; they invited the tiny diners in and used them as a delivery service. Some of the many spores on which the insects were feeding would stick to the insects' bodies and, when they visited another flower, the spores would fall off onto the female reproductive organ. In other words, the relationship between plants and insect pollinators is much more ancient than flowering plants.

Nevertheless, flowers changed the pollination game. Exploiting insects' keen vision, the first flowering plants used the bright color of their petals to advertise the precise location of their protein-rich pollen granules, which contain the male gametes. Before long, many flowering plants added in delicious scents and, best of all, sugary nectar to attract the insects. Keep each nectar snack small, and the insects work all day to get tiny mouthfuls of sugar from one flower after another, inadvertently transporting pollen among the flowers. Various plants changed the paradigm to attract birds (bright red flowers that are readily visible to birds but beyond the visual range of most insects) or bats (large, pale flowers that bloom at night). As a result, our world is filled with an amazing diversity of flowers and an equally amazing diversity of insects and other animals that earn their daily bread by consorting with flowers.

Then humans happened. As I've already written, humans love flowers. Ethnobotanists and archaeologists have shown that the pleasure of flowers is a constant across many cultures and over the millennia, indicating that flower appreciation is hardwired into our brains. The influence of flowers on our emotional state explains some of the ways we interact with flowers. Humans invest inordinate resources into constructing and maintaining ornamental gardens. Some gardens are famous, such as the legendary Hanging Gardens of Babylon built by the Assyrian King Nebuchadnezzar II for his wife (2600 BCE), the formal gardens of Versailles constructed for King Louis XIV of France (1661), and Frederick Law Olmsted's naturalistic landscapes of New York City's Central Park (1858), but most flower gardens are small-scale, amateur, and constructed in backyards. Humans also value floristry, the art of creating beautiful arrangements of cut flowers. Posies, bouquets, and floral arrangements in vases have been popular from the days of ancient Egypt (2500 BCE) to today. In some cultures, cut flowers are strung together to create elegant designs, as in the Hawaiian lei of plumeria or orchid flowers and the garlands of roses, violets, and lilies that decorated the walls of houses in ancient Rome.

Humans do not only use flowers to promote their individual sense of well-being. They also use flowers to communicate with other people. Flowers convey joy at a wedding, romantic love on Valentine's Day, sympathy at a funeral, apologies for something badly done, and so on. A gift of flowers can say, "My heart goes out to you in your troubles," "I love you," or "I am so sorry; please forgive me." Say it with flowers.

The outsized influence of flowers on our lives comes, in large part, from our influence on flowers. By selective breeding, we have created flowers that are larger, more colorful, more complex, and more scented, and so pack a bigger punch on our emotions. For millennia, we have been busy transforming flowers that were designed to manipulate pollinators into flowers that are emotionally pleasing to humans. For example, today's myriad varieties of ornamental roses are founded on a complex history of cultivation and crossbreeding among multiple wild rose species that stretches back more than 2,000 years in West Asia and Europe—and 5,000 years in China. In the New World, ornamental dahlias and marigolds, including double-flowered varieties (with multiple layers of petals), were developed by the Aztecs in Mexico more than two centuries before the Spanish conquest of 1521.

Flowers also feed our fascination with novelty. Something that is different is exciting and emotionally rewarding, and flowers from different parts of the world are perceived as particularly exotic. The trade in flowering plants in recent centuries has been so extensive that every gardening book and online resource is a cosmopolitan collage of modified plant life. We have become so familiar with garden plants that we need to remind ourselves that our backyards are a haven for the exotic. What's more, some of these exotic plants ignore the boundaries of our gardens and spread uncontrolled across the land. The most pernicious invasive plants include species that are courtesy of the gardening trade. Kudzu, tree of heaven, and giant hogweed come to mind.

I do hope that words can create a floral fix for a February day before the first flowers of spring. If not, it isn't so very long before the snowdrops, aconites, and crocuses brighten up our backyards; by April, we will be enjoying the real spring lilies in our forests, including the red and white trillium and the yellow trout lilies. If you can't wait, there is always the option of a visit to the local florist.

March

The calendar tells us that spring starts with the equinox on March 20. Too often, this is a cruel joke. March regularly brings days of bitter temperatures and yet more snowfall, along with other days of high winds and rainstorms that break up the ice on the lakes and ponds. Occasionally, we have a bright day with freezing temperatures only at night—the ideal conditions for collecting maple syrup. March sunshine, however brief, is always a reason to celebrate.

Whatever the vagaries of the weather, March is the month that starts in winter and ends with the first signs of spring. The first essay in this chapter recounts a chilly hike along Six Mile Creek, where the only color is found in the evergreen leaves of the magnificent eastern hemlock trees ("Hemlocks"). The next step is an equally wintry visit to Sapsucker Woods that inspired a reflection on woodpeckers ("Woodpeckers, Present and Absent"). As March progresses, many creatures are on the move. Mole salamanders make the arduous journey from their underground burrows to their breeding ponds ("Mole Salamanders"), and red-winged blackbirds are among the first of the summer visitors to return from their overwintering haunts in the south ("The Blackbirds Are Back").

Hemlocks

For me, the word *hemlock* always conjures up an image of a wooden cup. I mean the wooden cup right at the center of the enormous painting *The Death of Socrates*, painted by Jacques-Louis David in 1787 and now displayed in the Metropolitan Museum of Art in New York City. Socrates is accepting the cup into his right hand while looking the other way and pointing upward with his outstretched left hand. Apparently, he was thoroughly preoccupied by the philosophical implications of the possibility of an afterlife. The wooden cup is the infamous cup of hemlock, the standard method of execution in late fourth century BCE Athens.

The ancient Greeks could have used any part of the hemlock plant (*Conium maculatum*) to make their extract because the poisonous alkaloids permeate the entire plant, from the root tip to shoot, including petals and seeds. The officials responsible for carrying out the death sentence would have had no difficulty getting their supplies because hemlock is widespread throughout Western Europe, and it is very conspicuous. It grows to 5–8 feet and has clusters of many tiny white flowers and finely dissected leaves, like parsley. That is presumably why hemlock is sometimes referred to as poison parsley, although the nickname in Ireland, devil's porridge, is even more evocative. You cannot mistake a full-grown hemlock, with its purple stem and unpleasant, mousy smell. The catch is in early spring when the young shoots are still green and odor-free. Then, hemlock looks just like wild parsley . . . and there are a few fatalities every year. If you want food for free from the countryside, you need to know what you are doing.

Devil's porridge came to North America with the early European settlers, and it has spread to every US state. However, when people around here refer to hemlock, they don't mean *Conium maculatum*. They are thinking of a very different plant: the eastern hemlock, *Tsuga canadensis*. The eastern hemlock is routinely described as a graceful or handsome tree. In silhouette, it looks almost feathery because it sports masses of fine twigs, each one dipping gently to its tip. If you crush the leaf of the *Tsuga* hemlock, your fingers will smell kind of mousy, like the *Conium* hemlock. Consequently, the settlers gave the unfamiliar tree a familiar name: hemlock.

A quick leafy digression is in order here. Although the *Tsuga* hemlocks are in the pine family (i.e., they are related to spruces, pines, and firs), their needles are flat, not spiky, and very short. This makes them look almost like the *Thujas*, which also have flattened leaves but are members of the cypress family. Going further down the rabbit hole, don't forget that the cypresses include North American cedars, such as the majestic California redwoods (Sequoia sempervirens). These are not to be confused with old-world cedars (*Cedrus*, including the cedar of Lebanon), which have round needles and are in the pine family (i.e., they are fairly closely related to *Tsuga* hemlocks).

All of which is a wandering preamble to our recent walk along a stretch of Six Mile Creek. The water of the creek was churning, swollen by a partial snowmelt caused by two days of warm weather earlier in the week. For long stretches of our walk, the steep slopes that towered above us were clothed in maples, hickories, and oaks—leafless winter sentinels waiting for spring. In the hazy sunshine, it was a bright world of brilliant white snow and gray tree trunks. Then we entered a stand of hemlock trees. Under the dense canopy of evergreen leaves, this was a place of perpetual shade. It was noticeably cooler in the gloom. The snow lay thickly, and all sounds were muffled. We felt compelled to step lightly and walk in silence.

The local hemlock groves are very special. That makes it all the harder to reflect on the troubled history of this tree. The problem is that hemlock trees happen to be well endowed with tannins, which are complex polyphenols that precipitate proteins. These tannins deter most would-be munchers of hemlock leaves and bark; deer and other mammals, as well as most insects, leave these trees alone unless they are desperately hungry. It seems that the only animals that like tannins are humans, who use this material in traditional methods for tanning leather. The early colonists met their need for tannins by stripping bark from hemlock trees with a sharp knife, an activity that frequently killed the tree. The hemlock forests could tolerate small-scale, artisanal tanneries, but they were devastated when tanning grew to a major industry across upstate New York in the nineteenth century. Compounding this environmental vandalism, the tanneries were responsible for an uncontrolled release of toxic chemicals into the air and local waterways.

The hemlocks were saved by the discovery of synthetic tanning agents in the early twentieth century. The tanning industry was freed from the need for proximity to hemlock forests and moved to locations with better transport infrastructure. Over the last century, the hemlocks have gradually returned.

Now our hemlocks are facing a new danger: the hemlock woolly adelgid (*Adelges tsugae*). These aphid-like insects attach to the base of the needles and feed continuously on the plant sap. A heavy infestation weakens and, ultimately, kills the tree. The hemlock woolly adelgid probably arrived on a Japanese hemlock bought by a wealthy Virginia landowner who, in 1911, was constructing an oriental garden. Although the insects spread and were reported on hemlock trees in various places in Virginia by the 1950s, the hemlock woolly adelgid was not considered a serious pest—at least not until the 1980s, when they reached the hemlock forests of the Appalachian Mountains. Then all hell broke loose. Hemlocks, once widespread and common trees in the southern reaches of the Appalachians, suffered mass die-offs. Since then, the hemlock woolly adelgid has rapidly extended its range, including to New York State.

Nevertheless, we have reason to be optimistic for the hemlocks in our area. Entomologists have been investigating various natural enemies of the hemlock woolly adelgid. A contender for the role as the savior of the hemlock is *Laricobius nigrinus*. This small beetle hails from the Pacific Northwest of America, where it feeds exclusively on the native adelgids that infest the western hemlock tree. Luckily, these insects are also happy to eat the hemlock woolly adelgid introduced from Japan to the eastern US. *L. nigrinus* has been released into the hemlock forests of New York State and other states, down to Virginia and Georgia. It appears that the introduced beetles are getting established, and early data suggest they are reducing the adelgid populations.

The hemlock woolly adelgid is relatively easy to spot because the insects aggregate together and secrete lots of stringy, white wax that looks like an irregular ball of wool. These wool balls are evident most of the year, apart from high summer. As I have said, though, the trees by the path along Six Mile Creek looked healthy. We could see no tell-tale patches of white wool on the twigs or leaves. All good, so far.

Although my first thought on hearing the word *hemlock* is Socrates and his wooden cup of hemlock poison, my subsequent thoughts are always with the *Tsuga* hemlocks and the little brown beetles that may be their salvation.

Woodpeckers, Present and Absent

In early March, Sapsucker Woods is one of the very best birdy haunts in our local area. The deciduous woodland, swamp forest, and large beaver pond are all the more special because its name, Sapsucker, is a reminder of one of the loveliest of our summer visitors.

The sapsucker, or more correctly the yellow-bellied sapsucker, is a migratory woodpecker. During the summer months, we often see this pretty, little bird with its distinctive black and white plumage and bright red patches at the top of its head and nape of its neck. The yellow-bellied bit is all too easy to miss; there's a faint tinge of off-white or pale lemon on its front. All this is special enough, but the feeding habits of the sapsucker are even more remarkable.

As the name suggests, the sapsucker feeds on the sap of trees. It drills a hole into a living tree to access the vascular tissues lying just below the bark, then it lets the sweet sap ooze out and licks it up as it flows. It's a bit like a cat lapping up milk, only smarter. The outer edge of the sapsucker tongue is lined with tiny hairs that pull the sap into the bird's mouth by capillary action.

The sapsucker has two kinds of sap to choose from: the sugary phloem sap just under the bark and, a little deeper, the watery xylem sap. During the summer, the birds choose the phloem sap. They drill shallow, vertical slits into the tree, usually in a straight line going up the trunk, and then they keep the sap flowing by regularly returning and reopening the wound, often making the slit bigger. It's said that the sapsucker may have an anticoagulant in its saliva, so the wounds don't reseal. These gashes in the tree are called sap wells. The birds also snack on the soft plant tissues under the bark—and insects, seeds, and fruit are also on their menu. Sapsuckers have been seen to dunk individual ants or other insects into the sap flow before feeding their young with the sweetened insect meat.

In the dregs of winter, it is fun to think of sapsuckers, although we won't be seeing them until late April at the earliest. They will arrive hungry from their long journey from Florida, the Caribbean, or Mexico, but the phloem sap won't be ready for them until after bud burst in mid-May. Until then, the birds focus on the rising sap in the xylem vessels. The xylem sap wells are small and round, just the right size for inserting their beak, and very different from the gaping wounds of phloem sap wells in the summer. A single bird can tend up to a dozen neat xylem sap wells in a horizontal line around the trunk.

So there were no sapsuckers at Sapsucker Woods today. Instead, there were other woodpecker treats. We heard the unmistakable drumming of a pileated woodpecker, which is as big as a crow and as black as pitch apart from some white striping around the head and neck, and a rich red crest. Our noisy pileated woodpecker had clearly found a perfect dead tree for doing the pileated equivalent of banging on a drum. The rapid staccato drumbeats, very fast and in bursts, resonated around the woodland. In case anyone needed reminding, "This is our territory—all ours!" To be clear, "our" means a firmly bonded pair who live together in a well-defined territory through the year. There is great equality of the sexes in the pileated woodpecker world. Both sexes drum and maintain the territory. The male drives out male interlopers, and the female chases out females. Pity the young birds, whose chance of getting a territory depends on an older bird falling off its metaphorical perch. When that happens, the incoming bird bonds with the surviving resident.

The pileated is the biggest woodpecker in our neck of the woods, and it is probably the biggest woodpecker in the country.

We cannot be sure, however, because of the controversial status of another species: the ivory-billed woodpecker, a denizen of old-growth forests farther south. All the pictures suggest that the ivory-billed is the most elegant of birds, black and white with a great red crest and long, heavy beak that looks like pure ivory (but isn't, of course; bird beaks are made of cross-linked keratin, like fingernails). How wonderful to watch this great bird, perhaps sweeping from tree to tree with its 30-inch wingspan or stripping bark from a tree trunk with its fearsome beak to access beetle larvae. Many say that the ivory-billed woodpecker is strictly past tense, along with the dodo and the passenger pigeon, but there are persistent claimed sightings in the swamp forests of Louisiana. Although the photos and videos are smudgy, they have been interpreted by some people to match the size, wing beat characteristics, and flight speed of the ivory-billed. I want to be persuaded by these Louisiana sightings—and be encouraged by whisperings of sightings in the remote south of Cuba—but the evidence is disconcertingly flaky. Although there have been no definitive sightings since 1944, the U.S. Fish and Wildlife Service (2023) has hesitated to rule the ivory-billed woodpecker as definitively extinct in the US.

Let's shift our focus to the west of Mexico, the habitat of yet another woodpecker, the imperial. Closely related and similar in appearance to the ivory-billed, the imperial woodpecker is the largest woodpecker species in the world . . . but past tense is probably appropriate here, too. The imperial lived in small groups in mature pine forests, and, by all accounts, the birds were noisy, conspicuous, and slow flying. They didn't stand a chance when the loggers came in and the locals got access to guns. The last documented evidence for the imperial was in 1956.

Thank goodness, the conservation status of the pileated woodpecker is described as least concern. The pileated gets along very well in just about any mature woodland that has plenty of dead trees. Sapsucker Woods is perfect. We stood silently in the snow as the drumming echoed through the trees.

Mole Salamanders

It is received wisdom that the weight-bearing fin is a clever trick. I am referring to the paired fins at the front end (the pectoral fins) and back end (the pelvic fins) of the body; the dorsal fin, anal fin, and tail fin aren't weight bearing. Some might say that clunky, weight-bearing fins have their drawbacks. You may be a bit clumsy in open water, lacking the finesse with steering and braking that is possible with fancy fin rays. It might be a good idea to avoid gymnastics competitions alongside butterfly fish on a coral reef and to ignore odious comparisons with tuna fish, those top oceanic athletes with rippling body wall muscles and gargantuan tail thrusts. Instead, reflect on how weight-bearing fins at the front and back enable you to walk around on the seabed instead of squirming on your belly and allow you to lift your head out of a shallow oxygen-deficient swamp to gulp some air. Best of all, you can march fearlessly into a new world called land.

Before you know where you are, you are an amphibian and can walk, hop, or jump around on land. Most likely, though, you won't travel far because you have to return to water in order to breed. (I write "most likely" because a few amphibians have devised ways to keep their eggs damp on land.) The best time to come back to water to breed is the spring. There's plenty of water after the snow melts, and, with the warmer temperatures, the bottom of the food chain is getting into gear. That means oxygenated waters, courtesy of algal photosynthesis, and plenty of little animals, such as rotifers, amphipods, and flatworms, are on the menu for your offspring.

It is no surprise, then, that very early spring is the time of the salamander migrations. I am referring especially to mole salamanders (Ambystomatidae). These wonderful creatures are only found in North America, and at least eighteen species in total are in our area; including two common species, the Jefferson salamander (*Ambystoma jeffersonianum*) and the spotted salamander (*Ambystoma maculatum*). They are called mole salamanders because they spend the daytime, and the winter, in burrows that they dig in the loose soil of the forest floor. Although their legs are short and not so strong, these creatures can scoop out a burrow in soil already disturbed by other creatures, from earthworms to deer mice, or by frost heave during the winter. Digging duty is much reduced for those that happen upon the winter burrow of a hibernating mammal. These lucky ones just crawl in and share the residence.

The spring mole salamander migration is a Lilliputian wonder. It is a trek of up to several hundred yards from a relatively dry patch of woodland down to a pond. This trek can start anytime from sunset and can take hours because a salamander's limbs are short and not designed for sprinting. Adult life distant from a pond is not an option.

The perfect place to see the mole salamander migration hereabouts is the Robert Trent Jones Golf Course at Cornell University, which sports ponds and fragments of woodland alongside the fairways and greens. The ponds are shallow and, importantly, don't have fish that eat the salamander eggs and larvae. You could say the golf course landscape is designed for losing golf balls and promoting mole salamander populations.

We had our first mild and rainy day early in the third week of March this year. We left the car on the side of the road and took the path to a long fairway of manicured grass. We could feel the frozen ground beneath the squelch and puddles of the day's rain, and the remaining patches of snow glistened in the light of our flashlights. We walked slowly and steadily, scanning the ground with the flashlights. Aha! There's one! The unmistakable brown-gray Jefferson salamander, about 6 inches long, its head raised and beady black eyes glistening in our lights. It was marching purposefully down the slope toward the pond, its body undulating and its long tail dragging along the ground behind. The Jeffersons will mate in the pond, and the females will lay small groups of eggs onto twigs or other vegetation submerged in the water. The larvae will hatch two weeks later and will remain in the pond, eating and growing, until midsummer. Then they will metamorphose into adult salamanders and return to the woodland for their adult life of burrowing.

There was no sign of the spotted salamanders on that first trip to the golf course. The evening was still too cold, hovering just above freezing. The following week offered more rain and temperatures in the low 40s (6–9°C). This was much more the thing for spotted salamanders. We arrived soon after sunset, and we paced the fairway and edge of the adjacent woodland. The snow had gone, and the ground was soggy. It was perfect.

Armies of marching spotted salamanders were on the move, each individual in its uniform of black or dark gray with two lines of bright yellow spots extending along the body. The spotteds are bigger and chunkier than the Jeffersons but, like the Jeffersons, they were unfazed by the sudden exposure to the bright lights. If an animal was walking, it just kept on walking, and if it was still, it remained still. As with the Jefferson, the spotted lays eggs close to the pond edge, and the larvae will emerge and metamorphose during the summer.

I am neglecting one aspect of the spring salamander migrations. We are not the only people who find them splendid. On the

Jefferson night, there were a few other groups of people, all visible by their flashlights. But the night of the spotted was warm, and it turned out to be The Night. There were two or three other cars on the road when we arrived. By the time we had left at about 8:30 p.m., the site was a mass of marching flashlights and loud hollering, and more than fifty cars were parked nose-to-tail beside the road, right to the far end. It felt as though half of Ithaca was there. By 10 p.m., it would be a human scrum. Everyone seemed to be careful where they stepped, but one could not help but be anxious for the salamanders. This local human disturbance is trivial compared to all other amphibian travails caused by climate change, disease, and habitat loss. Still, I wonder whether we humans should limit our enthusiasm for events like the Golf Course Salamander Migration.

The Blackbirds Are Back

On the last Sunday morning of March, I was walking around the Cornell Botanic Gardens when, all of a sudden, my ears were assaulted by an unholy din. It was a cacophony of bird noise. The racket came in two parts. The first was a single note, endlessly repeated: *check-check-check*. That's the call. The second was a bit more complicated: two swift, clear notes and then a buzz. Think of *conk-la-ree*, with a vibrating flourish on the *ree*, and you are on the way to learning what is euphemistically called the song. Make sure to get your *conk-la-ree* all done and dusted within one second, then take a swift breath and say it again . . . and again . . . and . . .

It was wonderful. The song hardly compares to, say, a nightingale's, but it is a sure sign of spring.

The *conk-la-ree* bird is our local blackbird—or, to be formal, the red-winged blackbird (*Agelaius phoeniceus*). The male is perfectly black, including its eyes, beak, and legs right down to the claws, except for a bright red patch bordered with yellow at the bend of the wings. A quick dive into birdy vocabulary informs us that the bend in the wing is the wrist, and the colorful patches are known as epaulets. When the male is displaying, he puffs up his epaulets, making him look super-fierce to competitors and super-alluring to females.

The din was made by vast numbers of male blackbirds in the trees and on the reeds surrounding the main pond in the botanic garden. This is a perfect habitat for blackbirds, which like to be close to water, whether it is the edge of a lake or pond or a roadside ditch. The winter ice on the pond was almost entirely thawed, with just small fragments remaining, and the large goldfish were gliding about in the dark water. No turtles yet, though.

I looked about carefully for the female blackbirds, which are a fair bit smaller than the males and streaky brown, little different from the plumage of juvenile birds. I didn't see any. I must have visited the botanic gardens during the week or so after the males return to their breeding grounds to set up their territories, but before the females arrive.

For the male blackbird, a great deal hangs on getting the real estate in place. Prime territory and the most dashing epaulets mean that multiple females will be interested. A superstar male can persuade up to fifteen females to nest in his territory. That sounds tremendous, except that he is then duty bound to contribute to the feeding of many families. A successful male blackbird must look the part, sing the part, and then spend his days catching damselflies, mayflies, soldier flies, and the occasional moth for the many offspring of his harem.

His workload will be all the greater if the nest of one or more of his various mates receives a visit from a brown-headed cowbird. Alas, blackbirds are favorite hosts for eggs dumped by the female cowbird, and they are always duped into feeding the ravenous cowbird nestling in preference to their own young. As I walked away from the crowd of singing male blackbirds, a male cowbird flew up from the ground. He looked very sleek, his body a glossy black and his head a rich chocolate brown. The cowbirds are also planning for their breeding season.

Not all male blackbirds manage to carve out a territory. Indeed, most second-year males (meaning the birds that were born in the previous year) have no place to call their own, and they hang around through the breeding season. These birds are known as floaters, and they like to sneak in and mate with females when the territory-holding males are looking the other way. A floater will sire few, if any, offspring, but he doesn't have to work hard to feed the young, whether blackbird or cowbird.

The many struggles of parenthood, and the risks of being cuckolded by a floater or parasitized by a cowbird, are all in the future. For now, the priority for every male blackbird is to sing his heart out—to establish his place in the world.

April

April brings two certainties: the time to file our US tax returns and a rapid change in the seasons. But April is capricious. Although winter is in retreat, the timing of its definitive downfall is far from certain. A warm, sunny day can be followed by a snowstorm, and overnight frosts are a persistent threat. As we progress through the month, the pendulum between winter and summer continues to swing but with an ever-increasing bias toward summer.

One cannot help but admire the plants and animals that invest in growing and reproducing during April, despite frequently unfavorable and occasionally atrocious weather conditions. Pride of place must go to the skunk cabbage, one of the first native flowers of the year ("The Skunk Cabbage Classic"). These remarkable plants are usually evident in late March, and the skunk cabbage show is at its peak in early April. A further sure sign of early spring is the mating call of small tree frogs informally known as the spring peepers. The chorus of these frogs at favored ponds can be deafening ("Spring Peepers"). By April, American robins have returned to backyards all over town from their overwintering haunts, either in the local woodlands or in the south of the state and beyond. When the male robins join the dawn chorus, we know that winter is losing the fight ("Robins"). This chapter rounds off with a second plant that flowers in the early spring ("Wild Ginger"). Although less celebrated than the skunk cabbage, wild ginger has a story to tell.

The Skunk Cabbage Classic

Today is a very special day for every runner in the district. The first Sunday of April is the day of the Skunk Cabbage Classic, the first race of the year for the Finger Lakes Runners Club and named to celebrate the first flower of spring. I don't mean the snowdrops, which are getting past their prime, or the crocuses, now in great profusion. Both of these plants come from Europe. No, the first native flower of our region is the skunk cabbage, *Symplocarpus foetidus*, which is found in damp, boggy places and in shallow standing water.

The skunk cabbage is an arum and, as with other arums, its flower business is complicated. It has lots of tiny flowers borne on a central spike, which is surrounded by a modified leaf. Botanists have special names for the flower spike, the spadix, and for the modified leaf, the spathe.

The spadix of the skunk cabbage is rather short, like a knob, and studded with the mass of tiny, green petalless flowers. The spathe (that's the modified leaf, remember) is deep purple, mottled with green or gray, and folded over the spadix like a hood. The skunk cabbage wouldn't win against the snowdrop or crocus in a beauty parade, but it wins big time in the special interest category.

The skunk cabbage is the first flower of spring because it has its own central heating system. This is thanks to the mitochondria, the so-called powerhouses of the cells; they break down sugar derivatives to make chemical energy that fuels growth. It's all a bit different for many of the mitochondria in the skunk cabbage spadix (the flower knob). These mitochondria are modified to make heat, not chemical energy. The spadix is like a red-hot poker that melts snow and ice. In the late winter world, where an air temperature of 39°F (4°C) is considered warm, the spadix can reach 95°F (35°C). The thick, fleshy spathe that grows around the spadix is insulating, so the spadix and its flowers are nestled in a warm, dark cavity.

That's not all. The spadix produces volatiles—not the sweet scent of some spring flowers but the stink of rotting meat. The volatiles waft out of the cavity in the warm air that escapes from the gap in the spathe. If you or I get downwind of this, we'd likely wrinkle our nose and move on. But the first flesh flies and carrion beetles of the year respond very differently. They fly with great enthusiasm up the plume of warm, stinky air to find food and a place to lay their eggs. In the warm cavity of the flower head, the insects scrabble about, becoming dusted in pollen, but find no rotting dead bodies. The pollen-laden insects leave and at least some of them come across another inviting plume of skunk cabbage odor . . . and deposit the pollen onto the flowers of the second plant. Pity the hungry blowfly (the females may also be feeling egg-bound) as their hopes of a juicy, partly decayed mouse or sparrow carcass are dashed, again and again. An even more unfortunate fate awaits some flies that fall prey to one of the spiders that often lie in wait within the warm cavity of the skunk cabbage.

So the skunk cabbage of today, last week, and next week is all about spadix and spathe. Then everything changes. The pollinated flowers swell up and turn into little balls. These are the fruit. They start green and then turn black or blue. By late summer, the entire spadix starts to disintegrate, and the berries separate out. It's all very yummy for the creatures in damp woodlands, from the squirrels and muskrats to the wood ducks. The tiny seeds go through the guts of these animals and, with luck, end up in a perfect, soggy spot to germinate.

That's not all that happens. As the pollinated flowers turn into berries, the spathe (the first leaf) disintegrates and disappears, only to be replaced by lots of bright green leaves. In May and June, our local wetlands are dominated by enormous skunk cabbage leaves that are easily two feet long by a foot or more wide. Then, usually in the space of a few weeks, these gigantic leaves die and disappear. They aren't at all fibrous, and they leave nothing behind.

You'd imagine that these lush green leaves would be breakfast, lunch, and dinner for every passing herbivore. Not at all. You rarely see a hole or bite mark. That's because the leaves indulge in some serious chemistry. They are chock-full of calcium oxalate crystals, the same poison as in rhubarb leaves. On top of that, the leaves emit a fetid (hence *Symplocarpus foetidus*) mix of nasty chemicals when damaged, telling the world that they are not good to eat. Apparently, a single mouthful of raw skunk cabbage leaf makes the tongue, gums, and throat burn and sting for about two weeks. It would be far worse if you swallow the mouthful instead of spitting it out.

I've been a little economical with the truth. I've been writing about the chemical protection of the summer leaves. I should add that the same applies to the spathe and spadix of early spring. Every year, we marvel that these first flowers of spring are always perfect, untouched by hungry woodchucks, deer, and rabbits.

Except this year. Last weekend, we made our annual skunk cabbage pilgrimage to Sapsucker Woods. We walked along the boardwalk above the wetlands, confident that we were in prime skunk cabbage season. True. There they were in their thousands.

Surprisingly, though, many of them were damaged with part of the spathe eaten away and with the spadix dislodged or consumed. According to my friend Google, some animals are so desperately hungry in the early spring that they tolerate the nasty side effects of a skunk cabbage meal. Deer, beavers, black bears, and snapping turtles could be the culprit. I suspect beavers.

Getting back to the life cycle of the skunk cabbage plant, the next step is what happens after the leaves have died and the berries are eaten. Aboveground, the answer is an emphatic nothing. Underground is a very different story. The root system of every plant is enormous and keeps on growing. Furthermore, it is contractile, meaning the roots flex their proverbial muscles, pulling the root mass farther down, down, down into the goo of the bog. This protects the plant from being washed away during storms or snowmelt,. Skunk cabbages live for a good twenty years, digging themselves in more strongly year by year.

In other words, every time the runners in the Skunk Cabbage Classic flex their muscles, they are doing no more than mimicking the roots of the skunk cabbage.

Spring Peepers

The Oxford English Dictionary tells me that a peeper is something that peeps and that to peep is either "to look quickly and furtively" or "to make a weak or brief high-pitched sound." Peeping Toms and Easter chicks come to mind, but neither is correct. Spring peepers are frogs. Around here, spring peepers are the ultimate spring animals. The peeping starts in late March, some years at about the time of the salamander migrations (see March, "Mole Salamanders"), but the peepers come into their own during the opening weeks of April.

This year, an evening in the second week of April provided the perfect conditions for spring peepers. The wind had switched from bitter northerlies to balmy southerlies on the previous morning, giving us a warm sunny day. Then it rained with the wind direction barely altered. Just before 7 p.m., we bundled into the car and drove to Ringwood Ponds, a wonderful old-growth forest site with several deep kettle holes scoured out by glaciers, lots of swamp, and some vernal pools. Vernal pools are important habitats here. They are shallow depressions that fill with snowmelt water in the spring and then dry out, usually by midsummer. The big deal is that they are temporary, so they don't harbor hungry fish.

Our plan was to arrive at Ringwood Ponds about thirty minutes before sunset. That would mean we'd be able to enjoy the spring peepers as they gathered force from pianissimo in daylight to fortissimo after nightfall. We had come armed with flashlights and a map, determined not to get lost in the dark. We had decided to park in a pull-out on Ringwood Road near the entrance to the site. (This is an important detail.)

Before we reached the site, it was evident that our plan was in tatters. In full daylight, the spring peepers were calling, full blast. Hundreds and hundreds of male frogs were in position at the base of grass tussocks and bushes surrounding the pond nearest to the road. Each frog was belting out *peep-peep-peep*. . . . His only ambition in life is for his peeps to be louder and longer than the peeps of the neighboring frogs. That's the surest way to attract the attention of the females cruising around in the water or along the boggy edge. People say that the unsynchronized cacophony of peeps from so many frogs sounds like sleigh bells. I don't know about that. To me, they sound like spring peepers.

We had a second plan: walk cautiously toward the pond and watch the male peepers as they peeped. The loose skin around their throat blows out like a balloon . . . *peep* . . . and then slackens in preparation for the next peep. That plan crumbled, just like the first plan. The noise was deafening, and we would have damaged our ears if we had walked up close. So we didn't see a single spring peeper, which was a shame because they are very pretty, little frogs, a mere inch long and brown, with a dark X on their backs. Their Linnean name is *Pseudacris crucifer*: carrying the cross on their backs.

We ventured a little farther into the woodland, glad to put some distance between ourselves and the mega-decibel sleigh bells. Then we became aware of another sound, like quacking ducks. That was the wood frogs, *Rana sylvatica*. About double the size of spring peepers and another early breeder, the wood frog is routinely cited as the first frog of the spring. We walked carefully down to the shallow pond. It was now about sunset and no longer raining. Every ripple on the surface of the pond was caused by a male frog in search of females, which remained motionless in the water. You could say that the females were waiting to be caught.

The wood frogs weren't the only creatures in this pond. There were also lots of other salamander-like creatures, twisting and curling, their little legs flopping about as they moved. Even more of them were on the water's edge, making their way through the damp leaf litter toward the pond. In the fading light, the entire area was alive with the marching animals. Look out where you step!

We could not identify the marchers with any confidence. They were all darkly pigmented without any distinctive features evident in the rapidly dwindling light, and they were all smaller than the mole salamanders we had admired in March. This will make an interesting ID task for another year, we told ourselves.

It turned out we had come on precisely the right evening, probably one of the few Big Nights of the season—when vast numbers of early spring amphibians travel from their overwintering haunts to the ponds, where they mate and the females deposit their eggs. These amphibians overwinter on higher ground: spring peepers in leaf litter and behind loose tree bark, wood frogs in shallow burrows under leaf litter, and salamanders even deeper in the soil. The salamanders dig down below the frost line of the soil because they die if frozen, whereas the spring peepers and wood frogs can tolerate much of their body fluids freezing. The wood frog is so cold-tolerant that its range extends to north of the Arctic Circle.

For these early spring amphibians, the imperative is to get to the ponds as soon as possible after snowmelt. All that lovely water for their eggs and tadpoles . . . but the ponds will dry out before too long, and any offspring that hasn't metamorphosed into an adult in time will die. Despite this hazard, these vernal ponds are the right basket for the eggs because, as I mentioned earlier, they contain no hungry fish. Other species that come to breeding grounds later use persistent water bodies, and their offspring run the fishy gauntlet.

Ringwood Ponds would be the perfect place for these early spring amphibians, if it weren't for one thing. Recall that I mentioned how we parked along Ringwood Road? Most of our marching frogs and salamanders come down the hillside from their overwintering sites . . . and across Ringwood Road. Some years ago, herpetologists at Cornell University estimated that, on a Big Night, 20 percent of the individuals don't make it across the road. Alas, many frogs and salamanders stand stock still in the bright light of car headlights. The annual carnage is now much reduced, thanks to a "toad tunnel" constructed under the road. Unsightly but effective plastic fences extend from the tunnel entrance on each side of the road to guide the toads toward the tunnel. The animals use the tunnel to get to the ponds and, later, to return to the uplands. The toad tunnel is a welcome instance of ameliorating a problem that humans have made for other creatures.

We walked back to the car as night was falling and drove off along Ringwood Road—over the top of many more salamanders and frogs safely making their way through the tunnel to join the fun at the ponds.

Robins

A few years ago a small number of lesser celandine plants appeared under the maple tree in our backyard. We were delighted by the pretty, yellow flowers that attract lots of bees in the early spring sunshine. Since then, the lesser celandine has become the bane of our backyard lives. It has spread into all the flower beds, where it crowds out the emerging shoots of our herbaceous perennials, and onto the lawn where it threatens to choke the meadow violets.

The only way to stem the tide of the lesser celandine is to get out the trowel and dig it up. In this third week of April, I focused on the lawn. I started at the far end, treating every bright yellow flower as a flag for the trowel. I leave flowerless patches of leaves alone for fear that I might be targeting violets. (Although the leaves of the two species are very distinct in the flower book, the difference is much less clear cut in the real world.) The lawn is still wet and soggy, and every celandine excision results in a small divot. It's as if there's a third-rate golfer in the household. The damage is transient, though. The tangled bank of our back lawn soon covers the gap with a mix of bitter cress and chickweed, germinating from the seedbank, and the rapidly growing runners of white clover and wild strawberry. Every real gardener demanding a perfectly smooth velvet of green grass would be apoplectic about our lawn.

A character with a serious interest in our back lawn was scrutinizing—and applauding—my efforts. Throughout my celandine control session, he was dancing around just a few yards behind me. As we zigzagged across the lawn, I observed how he was checking out my last divot as I created the next one. I moved slowly and smoothly, watching from the corner of my eye. If my acknowledgment of his presence was too evident, he would fly off.

He is our backyard robin, and my divots disturbed small insects that provided him with one tasty snack after another. Such a familiar garden event conjures up the famous Beatrix Potter illustration of the robin perched on the handle of a spade, just next to Peter Rabbit munching radishes. But you need to de-conjure that image quickly because *Erithacus rubecula melophilus*, the much-loved and astonishingly tame robin redbreast of the UK, has never made it to the US. That's despite many efforts to introduce the species, partly because of the great affection early colonists had for the European robin and partly because it is mentioned by Shakespeare: "You have learned, like Sir Proteus, to wreathe your arms, like a malcontent; to relish a love-song, like a robin redbreast" (*The Two Gentlemen of Verona*, Act II, scene 1, lines 18–20).

In case you think that this digression is pure indulgence, please recall the far greater indulgence, and arguably insanity, of Eugene Schieffelin (1827–1906) and friends who invested vast amounts of time and money on a project to introduce all the birds mentioned in Shakespeare to North America. The most famous of the releases were made in New York City's Central Park.

My backyard robin is an American robin, named because it sports a rusty red breast that is sort-of-maybe like the European robin. In many ways, the American robin (*Turdus migratorius*) matches the Eurasian blackbird (*Turdus merula*) of the UK backyard. He works the lawn systematically, stopping here and there to thrust his beak into the soil for a worm or to snatch at small insects flushed up by his movement . . . or by my celandine control routine. The robin also enjoys slugs, snails, and, later in the year, berries galore.

However, an American robin is not a Eurasian blackbird with an orange chest. There are many important differences. For starters, Eurasian blackbirds are usually paired for life, whereas American robins are decidedly cavalier in their love life. The male robin that defends our backyard from all comers has only recently paired up with his lady love, and their relationship will end once the two broods are raised this summer. In November, the two parents will go their separate ways, and each will start afresh with a different partner next spring.

Another important difference is in their songs. We are all agreed that the Eurasian blackbird song is beautiful—a glorious contribution to the dawn chorus. My fellow citizens rave about the simple whistle of the American robin, but that's because they know no better. Biologists have noticed that the Eurasian blackbird song is what you would expect for a bird adapted to the high canopy of deep woodland. It is argued that blackbirds lived exclusively in wildwood of Britain until the latter part of the nineteenth century, when they exploded into suburban habitats. Consistent with this scenario, the blackbird became popular in British literature and poetry of the twentieth century (for example, Edward Thomas, R. S. Thomas), but earlier writers obsessed about the song thrush (for example, Thomas Hardy, Robert Browning, Alfred Tennyson). Shakespeare proves the point: the ousel cock (the blackbird) in Bottom's song is deep in the Forest of Arden, not on a palace lawn (*A Midsummer Night's Dream,* act 3, scene 1, line 127). Before you ask, the answer is *yes.* The Eurasian blackbird is another failed introduction to New York's Central Park.

Although the American robin's song is, in all honesty, rather boring, we welcome it as a sign of spring. It is special that, during April and into May, the robin is always the first to start singing in the morning, preceding the cardinal and well before sunrise. That is because the main job for the male American robin, for now, is to keep his territory intact. In our backyard, he sits around on the maple tree and the fence, he works the lawn, and he hoovers up the breadcrumbs we put on the deck for the juncos. He did little to help the missus as she built her nest in the box elder tree on the edge of our backyard, and now he is leaving her to incubate her eggs. Every now and again, she leaves the nest to get some food; she is reliably on the lawn with the male in the early evening. It's only after the eggs hatch that parental responsibility kicks in for the male. In about two weeks' time, we will see both birds foraging from dawn to dusk to feed their three to five offspring. Then the male will be on child-minding duty for the fledglings while the female sets up the next brood, usually in a new nest.

Provided a cat or blue jay doesn't intervene, we have lots of robin fun ahead of us in the coming months.

Wild Ginger

Robert H. Treman State Park is a superb place for woodland spring flowers. The paths follow about four miles of Enfield Creek at the bottom of a steep-sided gorge that is covered in hemlock, red oak, maples (red, sugar, striped), tuliptree, cucumber magnolia, and more. When we visited in mid-April, the *Hepatica*, among the first of the spring flowers, were already blooming and the *Trillium* lilies were in bud. We said that we must come back in a week's time for a wonderful *Trillium* show. We were true to our word, but, alas, the *Trillium* were no more advanced than in the previous week.

Strong westerlies and northwesterlies had brought a deep chill and snow from the Great Plains and Winnipeg for most of the week, culminating in a winter storm alert for Monday night through midday Tuesday. Five to nine inches of snow were predicted. In the end, it was just a couple of inches for us, but Binghamton, just forty miles southeast of us, got fourteen inches.

Our Treman walk on April 20 was not only a *Trillium* disappointment but also an occasion to celebrate a plant that is too easily overlooked. Close to the path at one location, there were a few heart-shaped leaves, all decidedly hairy. Look more carefully. Beneath the he leaves of each plant was a single rusty-brown flower, lying on its side and shaped like a jester's hat with three pointy extensions. This is the flower of the wild ginger. It is said in every guidebook that the flower looks and smells just like the thawing carcass of a wild animal that succumbed during the previous winter. Carrion flies are reported to scrabble about inside the jester's hat and pick up pollen, which they carry to another flower and deposit the pollen. Don't be too certain, though. The flowers are much less smelly than claimed (I can vouch for that), and these plants are mostly self-pollinated, with fly-mediated cross pollination a rare occurrence.

The Indigenous peoples of this area made good use of the wild ginger. They would dig up the swollen roots, dry them, and grind them to a powder—then add the powder to spice up a boring meal or boil freshly sliced roots in sugar water to make a spicy candy. The early colonists followed suit. During the twentieth century, chemists got to work, demonstrating that the roots contain aristolochic acid. "Great!" everyone said. Aristolochic acid, either synthetic or in wild ginger, was the perfect ingredient for various dietary supplements and a healthy flavoring for all sorts of medicines. But then the U.S. Food and Drug Administration (FDA) said, "Not so fast! We've just discovered that aristolochic acid is a carcinogen and causes kidney disease." The moral of this story is that wild ginger is best left undisturbed in the woodland. Keep your spade at home.

Let me quickly add that wild ginger is a very different plant from real ginger. Wild ginger, or more formally *Asarum canadense*, is a birthwort (family Aristolochiaceae) and related to magnolias. Real ginger is *Zingiber officionale* (in the family Zingiberaceae).

One more thing about birthworts: as well as wild ginger, there are various other birthworts native to North America, but, to my knowledge, there are no native birthworts in Britain. The one member of the birthwort family in British flower books is *Aristolochia clematitits*, informally referred to as birthwort and an ancient escape from the herb gardens of monasteries and abbeys. It was brought from its native range in southern Europe and prized for its medicinal properties (definitely pre-FDA ruling). In particular, birthwort tea induced abortions. Richard Mabey's *Flora Britannica* (1996) tells us that a prime site for finding it is among the nettles around the ruins of Godstow Nunnery near Oxford. He alludes to its particular value for the good ladies in the nunnery.

Real ginger is much more of a globe-trotter than birthwort. It has never been described in the wild, meaning that it is only known as a cultivated plant. Written records indicate that the Romans imported it from northeast India or southern China, where it is believed to have originated. The Romans didn't use ginger for cooking; they used it as a medicine to alleviate nausea. Apparently, the ginger trade was largely unaffected by the rise and fall of empires, Roman or otherwise. By medieval times, ginger was super-big business and increasingly used to flavor foods, especially breads, hence gingerbread. Over the last two millennia, India has been the main producer of ginger, but other countries got in on the act, including Thailand, Nigeria, and Jamaica. Ah, the sticky Jamaican ginger cake of my childhood in the UK! Now, one of my standards for afternoon tea is broonie, a US adaptation of Orkney gingerbread. The emphasis is on *adaptation* because it is sweetened with molasses, which would hardly figure large in the cuisine of the islands of Orkney, off the northern coast of Scotland.

Until very recently, it was far from clear where the real ginger plant came from in an evolutionary sense. The only certainty was that the first ginger lovers cultivated a wild plant, repeatedly selecting for the most fleshy and spicy rhizome. Perhaps they were like wine tasters, favoring the most luxuriant bouquet of ginger oils, including zingiberene, together with the lemon flavor of citral and a pungent mix of gingerols. Perhaps they loved the way that slices of the rhizome became sweeter and spicier when cooked—and more pungent when dried and powdered.

Recent molecular studies have given us some clues about the origins of culinary ginger. Its closest relative is the so-called beehive ginger *Zingiber spectabile*, a Malaysian species that is used as a medicinal herb and grown as an ornamental plant for its large red and yellow flower heads. The earliest ginger lovers' fixation on big rhizomes must have disfavored flowers because the ginger plant rarely flowers, and its flowers are small, pale yellow, and decidedly forgettable.

In the fascinating but complex world of gingers, two facts are indisputable: wild ginger isn't a real ginger, and real ginger isn't wild.

May

May is a glorious month. It is the first month of summer, although the locals love to recall the year when it snowed on Mother's Day and to predict the same for the current year. Despite the ever-present hazard of yet another wintry blast, this is the time when the trees come into leaf, and the forbs and grasses shoot up so fast that one can almost see them grow. Butterflies, bees, and other insects are on the wing, many sampling the nectar of early summer flowers. Everywhere, birds are singing at full volume, and many species are already fully engaged with the complexities of building nests and incubating eggs; by the end of the month, some are busy feeding their young.

The four essays for this month celebrate the lives of various vertebrate animals. I start with two regular arrivals to our backyard in early May: the chipping sparrow and the white-throated sparrow ("Hurrah for LBJs"). Both are unremarkable to look at—they are exemplars of what birders call "little brown jobs"—but their biology is very special. Then I take the opportunity to celebrate a mammal that is with us throughout the year but we rarely see: a porcupine, spotted in the local forest during the second week of May ("It's a Porcupine"). Onward to "Snakes"—creatures that spend the winter hidden away, usually in burrows below the frost line, and return to an active life aboveground in the spring. Finally, I circle back to birds. As I have already mentioned, the focus of most birds' lives in May is on their nests and offspring. As for any home, the maintenance and upkeep of a nest can be a complicated business ("Feather Your Nest").

Hurrah for LBJs

No, I don't mean hurrah for Lyndon B. Johnson, the thirty-sixth president of the United States. I mean hurrah for little brown jobs, the birder's term for those look-alike, drab bird species that include sparrows, buntings, some warblers and finches, and more. Please don't stop reading here. I promise you that LBJs can be interesting.

We have LBJs all year round. They skulk in the bushes, sit high in the trees, and putter around in long grass, always leading to our fragments of conversation about the streakiness of the breast, the presence or absence of a wing bar, the shape of the bill, and the length of the tail. But at this time of year, the diversity of LBJs sky-rockets. We are right in the middle of the spring migration. All those LBJs are on long journeys from their overwintering sites in the south (Florida, the Caribbean, and so on) to their breeding sites—perhaps here, perhaps in southern Canada, perhaps in the high Arctic. Species that would never live in our backyard (because it's the wrong latitude or habitat or because of the cat next door) drop down from the skies to rest and refuel for a few hours or even a day or two. Then they are gone.

With all this birdy excitement, the house is converted into a lux-ury bird hide, binoculars and bird book at the ready on the dining room table. My daily thirty-minute run up and down the drive (more like an ambling trot) and our daily tour of the backyard to admire the violets and red maple flowers may be postponed or even canceled because an LBJ is enjoying the backyard pitstop.

One of the many fun aspects of the spring migration is that the species vary from year to year. There's so much chance in which species happen to find our yard. In all this uncertainty, two species are regulars: the chippies and the whitethroats. Both are sparrows, meaning New World sparrows (family Passerellidae), which are not closely related to the Old World sparrows, such as the house sparrow. There are forty-two species of Passerellidae in the US, and all are LBJs.

Let's start with the chippy—or, more formally, the chipping sparrow (*Spizella passerina*). Over the last ten days, we have welcomed little flocks of migrating chippies to our backyard. About the same size as the European tree sparrow (which is smaller than the house sparrow) but more slender and with a longer tail, the chippy sports a beautiful rusty brown crown and perfect gray breast. Yes, we are in the LBJ world of brown and gray. The chippies call out *chip, chip, chip*, with no shift in pitch or intensity, as they forage on our unmown lawn, congregate in our maple tree, and then fly off.

We don't know where our here-today-gone-tomorrow chippies have come from. Officially, they overwinter in Florida and the Caribbean, as well as in California and Mexico, but some individuals haven't traveled that far because they hang around much farther north for part of or all the winter, varying from year to year. Their destination is similarly unknowable. Perhaps they will fly a few miles to a local breeding site, or perhaps they are gearing up for a long journey to Nova Scotia or the edge of Hudson Bay.

The chippy thrives in edge habitats, where woodland meets open lands. Before the Europeans, these habitats would have been rather few and far between, and, in all probability, the chippy was not a common bird. The haphazard clearing of the forests for small-scale farming by individual families and small communities was a bonanza for the chippy, and these birds are still very abundant in today's world of fragmented habitats, including many places (but not our backyard) in the Ithaca area.

The whitethroat, by which I mean the white-throated sparrow (*Zonotricha albicolis*), has also been much in evidence this spring. These birds are much chunkier than the chippies—and they're bigger, too. They hop around on the driveway, scratching about at the edge of the forsythia and spirea in search of insects and seeds. The big deal about whitethroats is their shiny white throat, bounded by a thin black line. That's rather bold for an LBJ, although it isn't a distinguishing feature. Some of the other forty-one New World sparrow species also sport white under the chin. Unlike the chippy, the white-throated sparrow is resident in a narrow sliver of latitude that includes parts of New York State: farther north a summer visitor, farther south a winter visitor. The birds that enjoy our backyard at this time of year are the migrants. Generally, the resident whitethroats stay in the forest through the year, and we rarely see them in town.

Whitethroats are very special for two reasons. The first is their song. It is astonishingly melodious for a sparrow, sounding like *Oh-sweet-Canada-Canada*, or so people say. Even if *sweet-Canada* is a bit of a stretch, the whitethroat song is in a different class from the metallic chips of the chippy. The migrants passing through have been regaling us over the last week or so.

The other special thing about whitethroats is that they come in two head colors. Bear with me . . . this LBJ story is really good. There's the white-striped and the tan-striped versions, meaning they sport a bright white stripe or a dull brown tan stripe running horizontally from just above the eye to the back of the head. Every year, we see the migrating white-striped white-throated sparrow in our backyard, but this year, for the first time, we have hosted a tan-striped, too. Drumrolls by lovers of LBJs!!

The white-striped is not more common than the tan-striped. In fact, they exist at almost a perfect 1:1 ratio for reasons that will soon be apparent. It's just that the white-striped sparrow is much bolder—more willing to show its face someplace that has a cat next door. Until the 1960s, it was believed that the tan-striped birds were juveniles, but then some LBJ fans got to work. Turns out: not at all! A white-throated sparrow is born with a head that is either white-striped or tan-striped, and there is no ducking that biological fate. All the females, whether they are white- or tan-striped adore tan-striped males, which are gentle, caring fathers that look after the babies. But white-striped females are much more assertive, and they get all the tan-striped guys, leaving the tan-striped females with the leftovers—meaning the aggressive, white-striped males. The males of the whitethroat world see it all differently. They all love the exciting, white-striped females, but, as I've already written, these females have eyes only for the tan-striped males. Therefore, the white-striped males make do with the boring, home-loving tan-striped females. Where studied, 95 percent of all pairs are between birds of different head stripes. Good thing, too! White–white pairings argue so much that they neglect the babies, and tan–tan pairings are so laid back that they are chased off their territory before the young are raised. Some people like to describe this setup as "the four sexes of the white-throated sparrow."

The chippy and whitethroat are just the start. Our bird list for the backyard and recent walks is bulging with LBJ sparrows, including the song sparrow, swamp sparrow, and white-crowned sparrow, as well as LBJ warblers, flycatchers, and finches.

If one is prepared to pay attention, this is the time of year to enjoy the LBJs in all their understated diversity.

It's a Porcupine

We had been chatting vaguely about visiting a preserve that is said to harbor a thriving population of porcupines. The chat was vague because the site is about an hour's drive from home, mostly on tedious, busy roads. There's also a low chance of seeing a porcupine at the otherwise unremarkable and rather small preserve. Instead, we were continuing with our weekly springtime visits to the local Robert H. Treman State Park. The scenery and waterfalls at Treman are spectacular, the walk is a good four miles and aerobic in places, and the spring flowers are glorious. Over the weeks, we have enjoyed the ever-changing displays of hepaticas, blue cohosh, lilies, Dutchman's-breeches, bloodroot, early saxifrage, trailing arbutus, spring beauty, and so on.

It was mid-May, and we were about a third of the way along the Treman walk. We were chatting about toothworts, the white-flowered crucifers that don't exactly make it to the premier league of showy flowers. We were busy concluding that they were a mix of the broad-leaved and cut-leaved, but not the slender, toothwort. Over this stretch of the walk, the gorge drops steeply from the path down to the creek at the bottom—so steeply that we looked straight into the top canopy of trees on the creek side of the path. Our toothwort conversation was halted midsentence by a dark brown, bushy structure on one of the top side branches of a hemlock tree. This strange thing was mostly round, but it was haloed in pale brown spikes and had an almost black extension pointing at about 45 degrees downward in our direction.

A quick check with the binoculars confirmed that we were observing a porcupine that was facing away from us, across to the far side of the gorge. The porcupine was perfectly still and balanced, apparently secure even though the branch was very slender and moving in the breeze. It was in the same spot when, later, we were retracing our steps back to trailhead. It hadn't moved at all. To the human eye, this was a decidedly precarious place to snooze the day away, but clearly the porcupine saw the world differently.

A good place to start describing the porcupine is its spiky halo of quills, distributed all over the back and sides of its body. The underside and head are quill-free, with just soft brown fur. Porcupine quills are modified hairs and adorned with many tiny barbs that point backward. If we had disturbed our porcupine, it would have raised each of its 30,000 quills by contracting special quill muscles in its skin. The mechanism is comparable to goose bumps in humans. The standard advice for anyone facing a porcupine with quill goose bumps is to back off fast. If you, your dog, or the local coyote is stupid enough to make contact, the quills are shed into the attacker's flesh, stuck firm with the tiny barbs. If the porcupine gets seriously upset, it flicks out its tail, driving its battery of tail quills, each covered in grease, deep into the flesh of your hand, your dog's nose, or the coyote's tongue. The winces and wriggles of the distraught enemy drive the well-greased quills ever deeper into the body, sometimes lacerating vital organs.

Altogether, the porcupine can lead a slow, easy life because it is rarely bothered by predators—and because it is a strict vegetarian. These solitary animals like to eat tree buds in spring, summer leaves, acorns and beech masts in the autumn, and small twigs and soft tissue under tree bark in the winter. For a porcupine, there's always plenty of food that doesn't run away.

Porcupines do have one problem, though; they are strapped for sodium. Most plant food has almost no sodium, but, as for all animals including humans, a porcupine's cells run on sodium. ('The biochemists in our midst will say, "Ah, the all-important

Na+/K+-ATPase.") This is the reason why porcupines occasionally take a swim in search of the sodium-rich rhizomes of the yellow waterlily, an abundant native pond plant in our region. For better or worse, humans have introduced other sources of sodium that are easier to access. For example, porcupines favor vegetation beside roads that are heavily salted in the winter, which leads to porcupine road fatalities. They also like to chew on sodium-rich synthetic rubber and structures made of plywood, which contains sodium-rich glue. The occasional reports of car tires completely chewed out and outhouses severely damaged in a single night are not exaggerations.

Let's hope that our porcupine on the gently rocking hemlock branch was dreaming of waterlily rhizomes and not car tires. If it was a female, it may have been thinking of motherhood. Porcupines mate in the fall, and the gestation period is a full seven months. The single offspring is born in May, emerging head first and fully furred but still enclosed in the amniotic sac because its quills would otherwise hurt the mother. Mom then eats the sac, and, within the hour, the baby's quills are hard and strong, ensuring that it is well protected from the start.

I will finish with a brief note of explanation. I am writing about one species, *Erethizon dorsatum*, of New World porcupines. This group originated in South America, and the ancestors of this one species invaded North America when the two continents collided more than two million years ago, an event known as the Great American Interchange. New World porcupines are different beasts from their Old World counterparts, which are bigger, don't like climbing trees, and live in Africa and parts of Asia. The Old World crested porcupine is also found in Italy, descendants of the species introduced by the Romans more than 2,000 years ago.

I find it rather amazing that two different groups of rodents hit on the same brilliant idea of quills independently, even though they lead their lives rather differently in other ways.

Snakes

We had splendid views of two species of snake during a single week in mid-May: the common garter snake and the northern water snake. To be precise, we had splendid views of mating snakes. As every snake knows, the first thing to do after emerging from a hibernating den is to procreate. Snake mating is a gymnastic affair involving intimate intertwining and thrashing about. The frenzy is induced by the female's perfume (pheromone), which is irresistible to any nearby male. Just one male had found our female water snake, so that was a fairly restrained affair. However, there were two males fighting for access to the female cloaca during our garter snake event. Premium snaky perfume can bring in a dozen or more males. Biologists like to describe the resultant scrum as a mating ball.

Let's start with the garter snake. It is by far the most common snake around here. When I am running on the driveway or weeding in the backyard in the summer, I regularly see one slither away. The garter snakes particularly like the flower bed by the east fence, a spot that is warmed by the sun most of the day. This species also lives very contentedly in meadows and woodlands, so long as it's not too dry. In fact, we spotted the mating trio this week in the leaf litter just beside a path in Robert H. Treman State Park. We were careful not to disturb them. That was more for their sakes than ours because garter snakes are not venomous, although a threatened garter snake will squirt out a stinky, greasy musk from glands at the base of the tail. The strenuously attended female garter snake at Treman will, in about ten weeks' time, give birth to a litter of up to forty baby snakes. That's the end of her responsibilities. Her babies are independent from birth. They have to work out how to catch insects, worms, and slugs on their own.

Our other snake of the week was the northern water snake, a substantially larger and stouter beast than the garter. I first became aware of the local water snakes about six weeks after I arrived in Ithaca. It was mid-September, and the summer lake-swimmers (humans, not snakes) were starting to return to the indoor swimming pool. Every year, lake swimmers return with stories of derring-do. It can be an encounter with a wayward boat dragging an anchor; becoming entangled in weeds, rising murk, or rip tides; coming face-to-face with a giant pike fish (with eyes as evil as a barracuda's); or getting bitten by a water snake. In my first year, I really wasn't sure how much of this to believe, but I decided on the spot that I was exclusively going to be a pool swimmer. So far, I have kept to that decision, but sometimes I wonder if I am missing out.

Yesterday evening, we went for drinks with friends who live by the lake. We were standing on the dock. As everyone chatted about this and that, I watched a magnificent five-foot-long water snake weave its way along the surface of the lake, its head raised like a periscope above the lapping waves. That was our second water snake sighting in the week. Our first sighting was a mating pair at the soggy edge of a water channel in the Binghamton University Nature Preserve. People were walking past, not noticing the writhing passion within six inches of the path. Perhaps that is for the best. The northern water snake can be very aggressive when disturbed. It will hiss, then flatten, then strike . . . and bite. The bite is painful but not life-threatening because the northern water snake has no venom. In any case, our pair of northerns did their thing undisturbed, and up to thirty baby snakes will be due sometime in August. Like the garter snake, the northern water snake gives birth to live young.

Several other snakes are quite common in our area. The milk snake deserves a special mention because it is big—all the better to hug its prey tight, like a boa constrictor. It is called a milk snake because people used to believe that it hangs around barns to suck milk from cows. Of course, it is only after the farmyard mice. The female milk snake lays eggs, usually a dozen or so, in the soil or in rotting wood and then slithers off. Her hatchlings emerge in late summer, and then it's their turn to slither off and do their independent thing. I should add that laying eggs is very much the standard way for snakes to reproduce. Getting pregnant, as with garter snakes and water snakes, is unusual for snakes.

Feather Your Nest

It is wise, at present, to sit at the far end of our deck table,—in other words, stay close to the house and not under the maple tree. Otherwise, you may find a strand of dead grass floating gently into your tea or a lump of wet moss landing with a thump on your plate. This is a temporary problem. Within a week or so, you may sit wherever you like.

The local difficulty on our deck is caused by a pair of robins building a nest in the fork between two branches of the red maple tree. When they started, the male came and went a few times with bits of grass or dirt in his beak, but he soon lost interest, and nest building appears to be the responsibility of the female. The outer wall of the nest is now in place. It is made of dead grass, bits of moss, and small twigs, and it looks a bit unkempt. Nevertheless, it is the inside that matters. The female robin works the nest materials into a smooth bowl by standing inside the nest and pressing firmly with the wrist of her wing. She may also add in some of her own feathers to keep it warm, along with some mud to make it firm, but I can't see that from below. Before long, the nest will be ready, and she will lay four bright blue eggs, one a day, and then incubate them for thirteen days. All being well, we will soon have a robin family above our deck table.

This pair of robins started family life rather late (see April, "Robins"). Perhaps they had previously tried somewhere else but found that their first choice was too windy or wet, or, more likely, the eggs were discovered by a chipmunk, squirrel, or blue jay. A different pair of robins has already built their nest in a crook of the box elder tree, and they are well on their way to parenthood.

Altogether, building a nest is a time-consuming and demanding business. You might expect that, once constructed, the robins would be house proud and do their due diligence to maintain it. Apparently not. In fact, they will likely move on to a different site and build a new nest for the second brood. This is partly to stay one step ahead of the chipmunk and company, but it is also good hygiene. A bird's nest is not only a cozy place to raise nestlings but also a warm café for parasites. Many of these parasites are equipped with skin-piercing jaws and feast on slurps of nestling blood. Some, such as louse flies, ticks, and bedbugs, wouldn't touch any food other than bird blood. Others, such as mites and fleas, are more catholic in their tastes, combining blood with feather dandruff, bird droppings, and the debris that inevitably accumulates when four tiny birds are living close together in a nest. Then there are the other insects and mites that feed on the parasites. In some ways, a bird's nest is an entire ecosystem.

It is also important for adult birds to practice good personal hygiene. Several times in recent days, I have seen the female robin take a break from building her nest in the maple tree. She flies down to the lawn, where she lowers herself into the ant colony by the washing line and remains completely motionless, her beak open. It looks like pure agony, and it probably is! All those aggressive ants squirting formic acid, which stings like crazy . . . but also kills off the mites, fleas, and ticks.

It is not just the robins that are busy keeping parasites at bay. As I was walking along the road yesterday, I saw a small group of house sparrows taking a dust bath. Their wings were flapping back and forth frantically, amid a cloud of dust that must have been two feet high. The tiny particles of sand and gravel clear their feathers of easily dislodged parasites and remove the dry skin and debris on which many parasites feed. A little distance from the bathing sparrows, there were two cowbirds, a male and a female, innocently sipping from a puddle of water. I wonder whether any of those dusted sparrows will be raising a baby cowbird this year.

Let's return to our deck. There is another nest close by. In recent years, a pair of starlings has nested in a cavity of our west neighbor's house, readily visible from our deck. Unlike the robins, which usually start afresh for each brood, starlings nest repeatedly in the same place, often after removing fouled nest material or even building a new nest on top of the old one. One morning this week, I watched five birds slip out of the hole—two sparkling black parents and three brownish offspring. They settled on the gutter and chattered, along with some whistles, then flew down to our damp lawn to hunt for worms. A starling nest is a much messier affair than a robin nest. The adults simply fill the cavity with grass, twigs, feathers, moss, fine bark, sometimes even bits of paper, cloth, and string, and then they mold out a depression in the middle for the eggs. We have also observed the parents flying to the nest with fresh green vegetation. It was impossible to identify the plants, but starlings are reported to favor wild carrot leaves and fleabanes, plants with an excellent reputation for suppressing mites and company. Despite these precautions, scientists who check starlings for parasites find that almost every bird bears ticks, mites, and lice. The lice are an Old World species, meaning that the first US starlings that were released in New York's Central Park brought their lice with them. . . . and the lice have hung on tight.

The starling family is decidedly noisy and messy. I strongly suspect that their days in our neighbor's house are numbered, and they will have to relocate. Perhaps they will leave many of their parasites behind when they set up elsewhere, building a new nest for their next brood.

June

The chief industry in Ithaca is education, and June marks a major shift in the academic cycle. The university and college students have departed, many directly following the commencement celebrations in the final weekend of May; and the staff and academics remaining on campus are no longer harried by crazy workloads and impossible deadlines. By the end of the third week of the month, K–12 schools close, releasing the students and teachers for the long summer vacation. Although it is business as usual for many people, our local world settles into a slower and more considered groove.

The natural world, similarly, gets into gear for the summer. Unlike the world of human affairs, though, much of the natural world is revved up, totally focused on the business of growing and reproducing. In most years, regular bouts of rain sustain the creeks and waterfalls and nourish the lush, green vegetation. As the days lengthen, little by little to the summer solstice, this world is bustling with life—especially bird and insect life.

Two of the essays for this month concern birds: the mockingbirds and their relatives, all great songsters ("Mockingbirds"), and the ospreys, which lord it over the Finger Lakes on long, narrow wings as they hunt for fish ("The Osprey"). Late June is also the time when caterpillars of the spongy moth appear; in some years, they cause devastating defoliation of oaks and other deciduous trees in the forest ("Spongy Trouble"). We begin the month, though, with an essay about flowers, specifically poppies, with some thoughts on flower names and their confusions ("Poppies").

Poppies

Our backyard is ablaze with the brilliant yellow of greater celandines. Several years ago, a greater celandine plant arrived under the black walnut tree, and we let it be. Our uninvited guest interpreted our response as an invitation to grow and multiply. We are entirely content with this arrangement. We and the bumblebees love the brilliant yellow flowers.

The greater celandine came to North America with Europeans, as a valued medicinal plant. The yellow latex that exudes from a broken stem was used as a remedy for jaundice and to treat warts. Rather weirdly, some people believed that swallows collect the latex and apply it to the eyes of their fledglings to help them see, so the greater celandine also acquired the rather dubious reputation as an eye medicine. Please don't try this out; the latex of greater celandine stems is highly corrosive.

Greater celandine is a stupid name for this lovely plant because it encourages a mental association with the lesser celandine. This is wrong because the greater celandine is a poppy (that latex is a clue) and the lesser celandine is a buttercup.

Last week, a friend came over for afternoon tea on the deck. She was delighted to see our greater celandine, but she knows it by a different name. She and other upstate New Yorkers call it "mustards" because the stem juice stains your fingers "as yellow as mustard." Then, barely drawing breath, she contrasted the welcome mustards with the disliked garlic mustard, which takes over any backyard unless weeded out assiduously. With my zero tolerance policy toward the highly invasive garlic mustard, I was totally in tune with our guest's sentiments . . . except that the word *mustard* lumps our pretty yellow poppy with garlic mustard and other real mustards, meaning the brassicas or crucifers. Our friend's "mustards" is no more a mustard than a buttercup. Naming stuff helps us make sense of the world, except that the sense-making can occasionally be nonsense.

Another plant is holding its own among the greater celandines under our black walnut tree. It is the Asian bleeding heart, *Lamprocapnos spectabilis*, a garden plant from China. I have read repeatedly that bleeding heart is a wonderfully descriptive name for *L. spectabilis* and species of the closely related genus *Dicentra*, because each flower is the color and shape of a rose-red heart with an extension at the bottom that looks like a drop of blood. This description sort of works for the fringing bleeding heart (*Dicentra eximia*), which is entirely red and native to the eastern US, but it's a stretch for the Asia bleeding heart in our backyard because its drop of blood is a brilliant white. Even less plausible are the two bleeding heart species that are abundant in our local woodlands. Both have entirely white flowers without a speck of red. One is called squirrel corn (*Dicentra canadensis*) because its underground storage organ both looks like the kernel of corn and is apparently eaten by squirrels and other small mammals. The other is Dutchman's-breeches (*Dicentra cucillaria*), named because the flower is shaped more like a Dutchman's pantaloons than a heart. Nevertheless, they are all bleeding hearts.

Our nomenclatural struggles with the bleeding hearts are not yet over. Botanists agree that bleeding hearts are closely related to poppies, including the greater celandine/mustards plant. Some say that bleeding hearts are a different group from poppies; others argue that they are a quirky kind of poppy. For now, the poppy people are winning. This means that I can say we have two kinds of poppy under our black walnut tree—the greater celandine and a bleeding heart—and neither looks like a poppy.

There are no "typical poppies" with bright red flowers that are native to North America, but the common poppy (*Papaver rhoeas*) was brought from Europe by accident. This poppy has done very well on the coattails of agricultural humans—so much so that no one knows exactly how or where it lived before people created fields of wheat and barley in the eastern Mediterranean ten thousand years ago. The poppy traveled to the British Isles with Neolithic farmers about six thousand years ago, and it arrived in North America with British colonists some four hundred years ago. Curiously, one of the several US names for the common poppy is the Shirley poppy. I suspect that is because the most popular garden variety of the common poppy is known as Shirley Single Mixed, even though wild poppies in the US are predominantly agricultural escapes, not garden escapes.

The opium poppy (*Papaver somniferum*) provides another twist to the struggles with names in the poppy family. The opium poppy is grown for its opium, a potent alkaloid in the latex of the unripe fruit capsules. The latex seeps out when small cuts are made in the capsule, which is very convenient for harvesting. There are other reasons to grow the opium poppy. In Britain, it is valued as a pretty garden plant, and it poses no drug-related risks because the opium content of the unripe capsule is negligible in the cool British climate. The opium poppy is also the source of poppy seeds that are used in cooking. As the poppy seeds mature, the opium content declines, and the poppy seeds are harvested from the ripe fruit capsules. The varieties grown in Europe yield blue-black seeds used in breads and cakes, whereas the varieties grown in Pakistan, India, and Malaysia produce white seeds used in curries.

The poppy seeds I buy in the grocery store are imported. That's because it is illegal to grow opium poppies in this country for any reason at all. And it's not just illegal; this law is enforced. The Drug Enforcement Administration (DEA) famously raided Thomas Jefferson's historic house, Monticello, in 1987 and dug up the opium poppy plants that had been grown in the kitchen garden for two hundred years. I am not sure if the lesson there is that no one is above the law . . . or that the DEA takes its time enforcing the law.

Part of me thinks it is a great shame that the poppy grown as a garden plant and for poppy seeds is painted with the same nomenclatural brush as the poppy grown for opium. Another part of me thinks it is always useful to know where our food comes from. After all, very tiny amounts of the opium alkaloids persist in the ripe fruit capsule of the opium poppy. There's not enough for poppy seeds to be psychoactive, but there is enough to be detected by the latest analytical methods used in drug testing. If you are considering an occupation that involves regular testing—perhaps a professional athlete or a member of the US armed forces—I recommend that you don't indulge in any curries that use poppy seed paste or any salads with vinaigrette from poppy seed oil. For the same reasons, you will have to decline when I offer you a slice of my lemon and poppy seed bread.

Mockingbirds

My first encounter with mockingbirds was literary. I was captivated by a story that had all the right ingredients: the goodies and baddies were unambiguous, and the plot bounded along with lots of action. Just as importantly, it was deliciously incomprehensible at every level.

The cover of the paperback version of *To Kill a Mockingbird* in my family's bookcase was stained with an ugly brown ring. Clearly, it had served as a placemat for a sloppy cup of tea, and I suspect that was its sole use before I started to read. I recall my father asking me what it was about. I replied without a moment's hesitation, "Mockingbirds," anticipating that these birds would soon figure prominently in the plot. I had no idea what a mockingbird might be, but I had the mental image of the Mock Turtle from *Alice's Adventures in Wonderland* with wings stuck on its back. I was probably eight years old. I am sure that, if my father had known the story, he would have realized it was totally unsuitable and taken the book from me.

In case you don't know, *To Kill a Mockingbird* tells you nothing about mockingbirds except that they make music, don't nest in corncribs, and won't get shot at by Jem Finch.

However, if one waits long enough, some mysteries solve themselves. I am now very familiar with mockingbirds. Scout Finch and I have the same species in mind—the northern mockingbird—which thrives in New York, Alabama, and every other continental US state, as well as the southern reaches of Canada and much of Mexico. The mockingbird isn't a glamorous bird, unless you are into "gray above" and "whitish" below, but its character more than makes up for these deficiencies.

The mockingbird has an exceptionally long tail, which it raises as it runs along the ground in search of insects, and then it does a wing flash, meaning it stops and raises its wings to "full-up" in a series of jerky movements as if it were a poorly oiled mechanical toy. In the process, a large white patch on each wing is revealed. Birders argue about why mockingbirds perform wing flashes.

Perhaps this behavior scares off predators, perhaps it scares insect prey out of the grass.

Mockingbirds also like to sit on a fence, at the top of a tree, or on a telegraph wire, where they sing gloriously. Of course, lots of birds sing, but the mockingbird is one of the few to make real music. The song is mostly whistles and trills, but the exact pattern of notes changes all the time. That's because mockingbirds keep on learning new songs, and they like to mimic the sounds of other birds, frogs, and even mechanical gadgets.

One small disappointment is that we don't often see or hear a mockingbird in our backyard. Our compensation is that a related species graces our backyard all summer. That's the gray catbird, whose song is just as varied but perhaps a little less fruity than the mockingbird's. In some ways, the catbird is more fun because it also mews like a cat, sometimes sounding more like a cat than a cat does! The catbird is a darker gray than the mockingbird, and it has a dainty black cap and, of course, a long tail. This year, we have a pair of catbirds nesting in the forsythia.

There's a third species of mockingbird found in the US: the brown thrasher. This bird is often cited in the books as prone to hiding under bushes and in woodland thickets. Two brown thrashers visited our backyard over a couple of days in late April this year, and I can vouch that they are very striking birds. Forget the shades-of-gray search image for the mockingbird and catbird. The brown thrasher is a rich chestnut brown above and has bold streaky spots on its creamy front. The thrasher visitation was a backyard first for us. We think these birds were taking a brief rest in our yard during their long migration north. As well as being a little higher on the glamour scale than the catbird or mockingbird, the brown thrasher is a musical wonder, reportedly with more than a thousand songs in its repertoire.

Altogether, mockingbirds and their friends—or, more formally, the Mimidae—are very special birds. They are restricted to the Americas. Other than mockingbirds, thrashers, and New World catbirds, there's one other type of mimid—tremblers, restricted to the Lesser Antilles in the Caribbean, from Saint Kitts to Grenada. Why are they called tremblers? When they wing flash, their entire body trembles.

The astonishingly musical songs of the mimids have been studied by researchers who have discovered that these birds have strict rules for how they put their songs together, and the rules are much the same as those used by human composers. Whether you are a mockingbird or a catbird, Mozart or Debussy, you organize notes into groups that go well together, change pitch and tempo gradually, and put notes of different timbre next to each other.

Various composers have been inspired by birdsong. For one composer, this inspiration was very personal. On May 27, 1784, Mozart bought a starling from a pet shop up the road from his house in Vienna. He was enthralled by one of the starling's songs, which included a seventeen-note melody—so much so that he used it as the theme for the final movement of the piano concerto he was composing at the time (Piano Concerto No. 17 in G, K. 453). Actually, Mozart changed one note, the ninth note, from G sharp to G natural. To my ear, his modification is more melodious and less starling.

It appears that Mozart was very fond of his pet starling. Alas, the starling died three years later, on June 4, 1787. Mozart's very next composition, completed on June 14, was . . . weird. The twenty-minute piece for strings and two horns is called *Ein musikalischer Spass* (K. 522), which means "a musical joke." It is full of strange discords and, in places, a mocking bass line that recalls

Shostakovich. Some people say it is a parody of bad composing, others say it is a memorial for his beloved starling, and a few say it is perfect demonstration of polytonality (music that uses two or more keys at the same time) long before its time.

My digression to Mozart and his starling is not as unrelated to mockingbirds as it may appear. The sister group of the Mimidae is the Sturnidae, meaning that the mockingbirds and starlings are cousins. Just imagine what Mozart might have composed if he had lived alongside mockingbirds, catbirds, and thrashers.

The Osprey

The osprey is a magnificent bird whose story is inextricably bound up with humans. In short, the osprey has been at the receiving end of the best and worst of human behavior, and its tale is one of survival, often against the odds.

Ospreys are one of the greatest pleasures of summer in the Finger Lakes. They are a regular, almost routine, sighting near Cayuga Lake. You can see them flying steadily over the water, great birds, nearly two feet long and with a wingspan approaching six feet, and almost entirely white from below, apart from a dark patch near the wrist of the wing. Keep your eyes open, and you may also see the massive hulk of an osprey surveying the world from the branch of a dead tree near the water's edge. In this posture, the dark brown upper side of the folded wings obscures much of the white underparts, although the white head with its broad brown eye stripe is obvious. Check out one of the many osprey nests and you'll see the naked heads of the youngsters bobbing up and down, especially when a parent returns with food. There's only fish on the menu because that's all ospreys eat. Every meal is delivered by foot, not by beak, because ospreys capture fish in their talons. When an osprey spots a fish from a considerable height, courtesy of its superb eyesight, it dives, free fall at breakneck speed, and plunges feet first into the water. There's no way the unsuspecting fish will slip away, thanks to spiky pads on the bottom of the osprey's feet and the double-jointed outer toes for extra maneuverability of the fearsome talons.

You don't have to go to an out-of-the-way place and sit huddled in a dark, claustrophobic bird hide to watch ospreys. These birds are not much bothered by human activity. One of the most accessible places to see ospreys in Ithaca is the Allan H. Treman State Marine Park (marine means water, including freshwater, here), which boasts a dog exercise area, a marina with picnic areas, tennis courts, an outdoor swimming pool, a playground, and a skating rink in the adjoining Cass Park. As I said, ospreys are not fussed by proximity to people.

All the same, we are lucky that the ospreys are here. We are at the southern limit of their summer range. If you look at a distribution map, you see that ospreys breed in the Finger Lakes and Adirondacks, around the Great Lakes, and northward through much of Canada. The osprey is resident in Florida and along the coast of Alabama and Mississippi and up the eastern seaboard to Virginia, but our birds don't come from there. No, our birds spend the winter in Central and South America. All this means that, for most of the US, ospreys are only seen during the spring and fall migrations. Yes, we are very lucky.

The osprey is found on every continent except Antarctica. In the Old World, most of the populations overwinter in Sub-Saharan Africa and eastward to India and Malaysia, and the birds migrate north to Russia, Scandinavia, and a few spots in Western Europe for the summer. There are also resident populations in the

Mediterranean and around Australia. Most birdy people say that all ospreys are a single species, but there are gainsayers who want to split the osprey into two, maybe even four, species that look the same and do the same thing but in different places. Everyone agrees that the osprey has no close relatives and that it is a distant cousin to eagles and hawks. No surprise there! However, the common-place reference of the osprey as a fish eagle or fish hawk is simply wrong.

I was surprised to read that the global conservation status of the osprey is least concern, with an estimated half-million individuals worldwide, although it remains a protected bird in New York State. It hasn't always been like this, especially in North America, where for many years the osprey population was tiny. This was thanks to DDT and related persistent organochlorine pesticides that were sprayed indiscriminately in the 1950s and 60s (see January, "With Fear and Trembling"). As we all know, these chemicals became progressively more concentrated going up the food chain, reaching dangerously high levels in top predator species, such as eagles, hawks, and ospreys. The North American osprey population has increased more than tenfold since DDT was banned in the 1970s, and the European populations tell a similar story.

The osprey has had a tough time in North America, but it was far worse in Europe. Long before DDT, the osprey was driven to extinction in the British Isles by a combination of trophy hunting and egg collecting. In Victorian Britain, no one would say no to a stuffed osprey in a glass case in their living room or to an osprey egg displayed in a specially constructed cabinet. There was much the same stupidity in North America, but fewer people and more ospreys meant this fashion had less awful consequences.

Today, the osprey population is increasing, thanks to greater environmental awareness and protection. We should never forget that routine sightings of ospreys throughout the summer are very special.

Spongy Trouble

Earlier this month, I went to the local garden supplies store for my annual purchase of a pheromone trap to control the Japanese beetles that infest our backyard. As always, I checked the store's whiteboard of seasonal tips. This time, it was all about the moth *Lymantria dispar*, recently given the name spongy moth (replacing the name gypsy moth and its derogatory connotations for the Romani people) (Entomological Society of America n.d.). Spongy moth was chosen because its egg masses look like a sponge and the moth is called *spongieuse* in French-speaking Canada and in France. The caterpillars of the spongy moth are avid munchers of the leaves of trees, and, in some years, they can cause serious damage—especially to their favorite, oak trees.

The following weekend, we visited Sweedler Nature Preserve, a steep gorge in mixed deciduous-hemlock woodland just beyond the northern end of Ithaca. It was a glorious sunny day, the sky a perfect bowl of blue, but as we stepped into the woodland, we could hear the steady patter of rain. How strange, especially as the canopy leaves were dappled in sunshine with clear blue above. It was also odd that the leaves of the high canopy were feathered, with some twigs completely leafless. Then we realized that it was raining caterpillar frass. An endless thrum of little black balls was falling to the ground all around us. The caterpillar rain accompanied us for the entire walk. What's more, the path was littered with leaf fragments, like green confetti. That's because the larger spongy moth caterpillars are not tidy eaters. Although these caterpillars prefer oak leaves, more than three hundred tree and shrub species can be on their menu.

Most of the trees at Sweedler Nature Preserve will survive the defoliation, and they will probably produce a flush of late-summer leaves after the caterpillars pupate, sometime in August. They will be weakened, though, making them more susceptible to disease or drought.

It was a few days later, while we were eating our breakfast on the deck under the maple tree, that we realized the caterpillar rain had come to us. Very quickly, we discovered that it is not a good idea to have cereal bowls or cups of coffee underneath a horde of pooping caterpillars. We solved the problem by migrating to the middle of the lawn for our meals. The more difficult issue was how to protect our trees from the spongy moth caterpillar.

I am undecided about whom to blame for this. Perhaps I should be holding Étienne Léopold Trouvelot to account—or Charles-Louis Napoléon Bonaparte. It was in 1851 that this particular Bonaparte decided he was fed up with being the president of France and turned himself into Emperor Napoleon III. This caused a spot of bother for Trouvelot, who wasn't just a proficient artist, an amateur astronomer, and an entomologist. He was also a vociferous republican, meaning that he abhorred any notion of emperors. It would be prudent to leave France until things cooled down a bit, so Trouvelot and his family went to the US and, by all reports, settled contentedly in Medford, Massachusetts, near Boston.

Trouvelot had a brilliant idea: develop a silk production business from silk-producing caterpillars. He focused mainly on native US silk moths, but there were two problems. The US species don't make much silk, and they are susceptible to bacterial infections when caged together in large numbers, as required for a silk factory. Trouvelot's big idea was to generate hybrids between wild silk moths and the naturally resistant spongy moth from France. His idea was crazy because wild silk moths are in the family Saturniidae, and the spongy moth is in the family Erebidae. You'd have more luck mating a lion and an elephant. Of course, it all came to nothing, and Trouvelot soon lost interest. Subsequently, he got a job as an astronomer at Harvard University, and he didn't look back. He authored many publications, especially on sunspots, and craters on both the moon and Mars have been named in his honor.

There is, however, a legacy to Trouvelot's silk project. Some of his spongy moths escaped into the woodlands of Massachusetts. Within ten years, the first defoliating outbreak was recorded, and there have been outbreaks every ten to fifteen years ever since. The spongy moth is now found throughout the northeast, south to North Carolina and west to Minnesota. Ecologists have studied the species intensively. It has a simple life cycle: the insects overwinter as eggs, which hatch into caterpillars in the spring. By August, the caterpillars are ready to develop into pupae, and the adult moths live for just a few weeks in late summer.

There remains the pressing issue of what to do about our spongy moth caterpillars. I thought about the seasonal tips whiteboard of the garden supplies store. At first sight, a sticky trap around the base of the tree sounded good. The larger caterpillars feed only during the night, and they descend to a protected spot near or at the bottom of the tree during the day. A sticky trap would get every passing caterpillar on its daily commute, but it would also catch everything else that crawled up and down the tree. The collateral damage would be enormous! The other two options on the whiteboard were organic insecticides, meaning that the products come from an organism and have not been synthesized by chemists. Both suggested insecticides come from bacteria: one a type of Bt toxin that kills any butterfly or moth and the other Spinosad, which kills any insect. That sounds like more killing fields.

I needed something specific. I turned to the internet and discovered the burlap trap. Burlap, also known as hessian sacking, is available in thirty-foot rolls from our garden supplies store. The

technology of the burlap trap is child's play. Wrap a length of burlap around the tree trunk and tie it around the middle with a piece
of string. Then flip the top half over. This creates a snug, dark place
for insects to hide between the two layers of burlap. Finally, visit
the burlap traps once a day with a bucket of soapy water and pick
off every foul, wretched spongy moth caterpillar that is sheltering
there. Leave the ants, beetles, earwigs, and other beasts that are
making use of the burlap shelter.

Within a week, we had our backyard spongy moth infestation
under control, and we created new daily entertainment of seeing all manner of other creatures making a home in the burlap. If
only low-tech strategies were available for managing all invasive
insect pests.

July

In our area, July is the peak month for vacations. From the Independence Day celebrations through the end of the month, campgrounds and cabins in the local forests and state parks are booked, public swimming sites are busy, and the aroma of barbeques hangs in the humid air of many evenings. Wild places offer a kaleidoscope of colors, as flowers advertise their nectar and pollen to bees, butterflies, hummingbirds, and other pollinators. Meanwhile, the birdsong that filled our world in May is largely gone, and we are left with the occasional burst of begging calls from nestlings hidden deep in a bush and companiable contact calls between mates or between parent and fledgling offspring. In the closing days of the month, a different category of sound starts up. It is the time for insect music makers. The first of the cicadas, together with the earliest of the crickets, grasshoppers, and katydids, are testing out their courtship songs.

Three of the essays this month focus on individual species: a strikingly beautiful butterfly that is on the wing for just a few weeks around midsummer ("The Baltimore Checkerspot"), the American toad with its voracious appetite ("A Natural Corridor for Toads"), and the American sycamore tree that inhabits the damp valley bottoms and has been planted in town ("Shedding

Bark"). The final essay was inspired by a visit to Green Lakes State Park, with its clear, green waters created by a fascinating interplay of biology and chemistry ("The Making of a Green Lake").

The Baltimore Checkerspot

The Baltimore checkerspot is a butterfly. It majors in orange, black, and white. The upper side of its wings is black with rows of white blobs, known as checkers, and there's bright orange both around the margin and as spots on the inner, black region. When the butterfly is at rest with its wings up, its underside gleams with white and orange checkers. Continuing the theme, the butterfly's body is a patchwork of white and orange checkers on black, its head and legs are orange, and its black antennae bear a flamboyant orange knob at the tip.

The Baltimore checkerspot does not get its name from the city of Baltimore but from its gaudy color scheme. This butterfly was formally identified in 1773 by Dru Drury, who lived in London, UK, and paid sailors to bring him insects from exotic lands, including North America. He likened the appearance of one of the dead butterflies that he received to the yellow and black coat of arms of George Calvert, the first Lord Baltimore, who was a renowned politician in the court of King James I of England. Not everything went well for George Calvert. He got into major difficulties in 1625 when he mishandled negotiations for James's son Charles to marry one of the Hapsburg princesses and then was outed as a Roman Catholic. This double whammy ruined his political ambitions. Thereafter, he campaigned for the settlement of persecuted British Catholics in North America, starting with the Colony of Avalon in Newfoundland. That was a bit chilly, so he set up a second colony in Maryland, and his son Leonard Calvert became its first colonial governor. In 1729, one of the towns there was named Baltimore in honor of Cecil Calvert, another of George's sons and the second Lord Baltimore. Perhaps it is fitting that Maryland has adopted the Baltimore checkerspot as its official insect.

Back to the "here" of Ithaca and the "now" of early July, several Baltimore checkerspot butterflies have been flitting about in a patch of damp meadowland at the southern edge of Cayuga Lake over the last few days. We were very lucky to see them because the Baltimore checkerspot flies for just three weeks, between late June and early July, and is not at all common.

The Baltimore checkerspot has just one generation per year. The adults emerge in late June, and the females lay batches of up to one thousand eggs on a suitable food plant for the caterpillars. Then the adults die. Most butterflies with a single generation per year overwinter as eggs or, alternatively, as pupae, but the Baltimore checkerspot isn't an ordinary butterfly. It spends nearly 85 percent of its life—forty-four weeks, including the entire winter—as a caterpillar. If you are like me and care about these things, it spends just three weeks as an egg, two weeks as a pupa, and three weeks as an adult.

Let's see how this works. Focus on one egg in a batch of, say, five hundred eggs, laid on the underside of a leaf of their preferred plant, the white turtlehead. Our egg hatches into a little caterpillar, along with its 499 siblings. It likes company, and the entire gang climbs to the top of the turtlehead. There they set about creating a communal tent of silk that encloses much or all of the plant. This is the so-called feeding web, within which the caterpillars munch steadily and grow and pass through two molts. As our caterpillar gets bigger, it may occasionally climb out of the tent and take a brief foray to feed on other plants, but it always comes back to the tent within a few hours. Then, sometime in August or September, all the caterpillars stop feeding and work together to create a thicker tent of silk, within which they do nothing. That's the pre-hibernation web.

It starts to get cold, and our lazy, do-nothing caterpillars get fidgety. All five hundred caterpillars climb out of their web and go down into the dying grass and leaf litter beside their turtle-head plant. After a few days, groups of them leave the big group and wrap themselves up in a communal tent that they make of plant debris, held together with their silk. Then they do a lot more sitting around, much of the time under a layer of snow. When spring arrives, they wriggle out of their tent, eat some food, grow a bit more, and complete their development to adults.

At first sight, it makes no sense for all these juicy caterpillars to sit around doing nothing. You'd imagine they would take pride of place on the menu of any bird, mouse, frog, beetle, or spider. However, the caterpillars are safe because they are bright yellow with black stripes and decorated with lots of thick black spines, informing every hungry beast that they are poisonous. Their color signal is entirely honest. Their tissues are laced with toxic iridoid glycosides, which they derive from their turtlehead food plant.

Still, the caterpillars are not without enemies, especially one tiny species of wasp called *Apanteles euphydryidis*. The females of this wasp have a single ambition in life: to squirt an egg through her sharp ovipositor into a juicy Baltimore checkerspot caterpillar. The wasp egg hatches and the larva munches steadily on the tissues of the caterpillar, which stays alive for a while; eventually, the cater-pillar dies, and the adult wasp emerges from the carcass. Protec-tion against these wasps is provided by the communal webs. If the caterpillars stay inside their tent, the wasps can't reach them. The wasps sit on the outside of the web, often in large numbers, and attack any caterpillar that sneaks out. Many wasp attacks are unsuccessful because the caterpillars can knock the wasps away with a jerk of the head, but some head jerks are not fast enough or strong enough.

The Baltimore checkerspot is one of many butterflies with declining populations because humans are destroying their preferred habitat—damp meadows—at an unprecedented rate. That said, I suspect this species has never been common because its preferred plant, the white turtlehead, is not abundant. Furthermore, the butterflies stay close to home. Once a butterfly has found a bunch of white turtleheads, it just stays put . . . and so do its children, and grandchildren, and so on. Some butterfly conservationists decided to help the Baltimore checkerspot by planting extra clumps of turtleheads, and it took the resident population five years to expand to use the new plants, just six hundred yards away.

Nevertheless, the prospects for the Baltimore checkerspot are far from gloomy. That's because of a remarkable change in the behavior of some individual females, which have taken to laying their eggs on English plantain (*Plantago lanceolata*), a widespread invasive plant that favors dry grasslands rather than the damp meadows where turtleheads thrive. The caterpillars grow and develop on the plantain, which, like the turtleheads, contain the iridoid glycosides the insects need for protection against predators. In some locations with abundant plantains, the populations of Baltimore checkerspots are increasing. It appears that this butterfly is adapting to some aspects of habitat change caused by humans.

A Natural Corridor for Toads

Corridors in buildings tend to be unattractive but necessary routes to facilitate access between different rooms. Natural corridors serve a similar purpose as connectors between patches of natural habitat that would otherwise be isolated by human activities, but they are often interesting places as wildlife hotspots in their own right.

One of the most varied local natural corridors is Monkey Run, a stretch of tree-lined banks along a portion of Fall Creek, just upstream from the main Cornell University campus. One moment, you are sauntering beside the creek in a floodplain forest of maple, sycamore, willow, and quaking aspen. Then, suddenly, you come up through a steep hemlock grove and emerge at the top of a vertical cliff in a dry oak-hickory forest with a fantastic view of the creek a hundred feet below you. But this isn't a standard Ithacan gorge of Devonian sedimentary rocks worn away by glaciers and rivers. Monkey Run's cliffs are crumbly, turn to mud in rainy weather, and are continuously eroding away. In some places, it is ill-advised to go to the cliff edge because you may hear a tiny rumble and then find that there is nothing beneath your feet. These cliffs are glacial till from the last ice age and known locally as the Varna Cliffs. That sounds exotic to me, like something out of *Star Wars*, but the cliffs are named after the nearby settlement of Varna, which boasts two auto repair shops and a launderette.

We were walking along a wet patch of the trail beneath a dense maple canopy, focused mainly on swatting away insects that were hungry for our blood and sweat. As we took the next step, there in front of us was a full-grown American toad, which hopped languidly to the undergrowth on the edge of the path. It is very easy to declare that American toads are no beauties. The words used to describe all toads instruct us to see them as ugly. Toads are squat; their movements are ungainly, especially compared to leaping frogs; and their dry, wrinkly skin is littered with bumps that are usually referred to as warts. American toads tick all these ugly boxes with gusto. They are large beasts—our toad must have been a good three inches long—and their warts are numerous and irregular in size and shape.

The edge of Fall Creek in Monkey Run is the perfect place for American toads. As our *National Audubon Society Field Guide to Reptiles and Amphibians* (Behler and King 1979, 387) informs us, this species favors places where "there are abundant insects and moisture." It has been estimated that a single toad consumes ten thousand insects in a single summer, although I struggle to imagine how that was calculated. The American toad is very common and widespread across the eastern part of the continent. What was special about our sighting was that one rarely sees these toads during the day. These animals tend to be nocturnal and spend the daytime under logs or big stones.

A little farther on, we saw three tiny toadlets walking along the path. I suspect they were newly minted American toads. The timing was right: the females deposit their egg strings, each containing thousands of eggs, into water in April, and the egg and tadpole stages generally last about two months or so. Our toadlets have a lot of growing to do, but they have time. American toads generally take two to three years to reach maturity. For better or worse, very few make it. Our toadlets have already survived the tadpole stage, which is much loved by hungry fish, birds, and water beetles—and toad, both large and small, is on the menu for every garter snake

This does not mean that the American toad is without defenses. Its oh-so-ugly skin is excellent camouflage . . . and the toad stage is poisonous. That is thanks to two bumps, one on either side, just behind the eyes. As we watched our toad, we saw the two bumps, both perfectly regular in shape and a uniform pale brown. They are the parotoid glands. A threatened American toad releases a toxic milky colored fluid from these glands. It is said that a dog that attacks an American toad will back off fast, foaming at the mouth and yelping, and will never make that mistake again. Garter snakes and other toad specialists have special detox enzymes that neutralize the toad's defense.

The American toad isn't the only beast with parotoid glands. These structures are found in a variety of toads, frogs, and salamanders. The poisons are generally alkaloids, and they do nasty things to the nervous system of unprotected animals.

Earlier, I explained that Monkey Run is a natural corridor. That's important. It means that the creek is bounded by a relatively narrow strip of woodland. I am confident that one of the reasons why it is such a good spot for nature is that this corridor is sandwiched between large expanses of unimproved meadows that are only intermittently cut back. The meadows are a buffer between the forest strip and destructive human activities, and the wildlife in all the habitats is enriched by the proximity to the other habitats.

The trail took us from the damp woodland home of the toads into the bright sunshine of one of the meadows. At this time of year, the vegetation is home to red-winged blackbirds, savannah sparrows, and song sparrows—all singing at top volume—and to butterflies galore. We spotted seven species, including monarchs sizing up the milkweeds for laying their eggs, the fast-flying tiger swallowtails, and clouded sulfurs. A little farther on, there was a small group of meadowlarks, small birds that sport a brilliant yellow front and a vivid black *V* across the chest. Unfortunately, meadowlark populations are in decline. Their habit of nesting on the ground in fields and grasslands makes them vulnerable, but I like to think they are safe in the old field habitats around Monkey Run.

What we need are more places like this: diverse and rich habitats where animals and plants can lead their lives without disturbance.

Shedding Bark

The American sycamore tree (*Platanus occidentalis*) is a highlight of our walks from the door. A regimented line of mature and decidedly gnarled specimens graces the side of a nearby street, and it is the dominant tree along the valley bottom of the nearby Six Mile Creek, where it thrives in natural disorder.

In the last few weeks, though, something strange has been happening to the sycamore trees. As we walk under them, whether it's along the creek footpath or down the neighboring road, the ground crunches. The crunching isn't gravel; it's lots of bark that has dropped off the trees in great chunks. The dark brown bark of the upper branches and, for some trees, the main trunk is peeling away, exposing an inner yellow or gray surface. Every now and again, these peelings drop to the ground with a thud. It almost looks like another dreaded disease, but it isn't. The bark shedding is healthy but, for some reason, excessive this summer.

I decided to look into why sycamore trees engage in the strange exfoliation of their bark. At first sight, it seems decidedly unclever. Bark is always cited as the tree's first line of defense. It's the outer skin of dead cells, like the outermost layer of human skin, and it serves much the same purpose as our skin does for us. It's a waterproof layer that keeps internal water in when it's dry and external water out when it's wet. It also guards against all but the most determined of pathogens, including bacteria and fungi, and against parasites, such as insects and worms. This protection is super important for the trunk and branches of a tree because its circulatory system is a narrow ring just under the bark.

All sorts of reasons have been proposed for why sycamore trees shed their bark. They seem to be cited as alternatives, but most of the explanations could be both-and rather than either-or. The first explanation is a bit like the reasons health spas give for subjecting their victims to facials. Exfoliation, meaning the removal of the outermost layer of dead cells, is meant to brighten you up. For a tree, the process removes all the algae and lichens that settle on the bark, along with the surface ecosystem of munching mites, insects, and worms. The argument that bark shedding is comparable to spring cleaning, brightening up, and "making a fresh start" sounds a bit like the advertising blurb for spa treatments to me, and it's no more convincing.

A variant of this explanation is that the smooth trunk and branches of trees that have shed much of their bark are slippery. This exfoliated surface offers poor purchase for caterpillars and other insects that clamber up, down, and around, and it provides no cracks or crevices in which they can hide. Furthermore, caterpillars, other insects, and insect eggs—on the sheer, pale surface of an exfoliated sycamore tree—would be obvious to hungry birds, spiders, and squirrels. This idea seems plausible, and I wonder if this might be one reason why the sycamore tree is not badly affected by caterpillars of the spongy moth.

The second explanation couldn't be more different. It holds that a tree can sometimes grow faster than its bark can extend. Let's unpack that. Trees are not the same as insects, which are stuck inside an inextensible suit of armor. The only way an insect can grow is to crack open its skeleton and turn into a puddle of tissue while a new, bigger skeleton is made. By contrast, trees can add more bark as their trunk and branches expand. However, if the tree grows very fast, then this incremental patch filling may be insufficient, and it is better to remove the corset (meaning the bark) and build a new, bigger corset (that is one size bigger) after the growth spurt. We've had a very favorable spring and early summer with extremes of neither temperature nor rainfall. Perhaps our sycamore trees are simply taking the corsets off.

Now for the variant of this explanation. Without the bark, the ring of living tissue under the old bark is exposed to the light. That's a great opportunity for some trunky or branchy photosynthesis to supplement the photosynthesis happening in the leaves. Perhaps our sycamore trees will need new corsets that are two sizes larger after this summer.

It is rather special that sycamore trees indulge in extensive bark shedding. Some trees, such as birches, do it in a small way and probably for similar reasons. Others, notably the shagbark hickory, hang onto their bark peelings as protection against fire.

I suspect that the sycamore is a supershedder because it grows fast and can get seriously big. I've read that the first European settlers in North America found enormously large and ancient sycamore trees in valley bottoms, so much so that some trees had naturally hollowed out on one side. Apparently, these settlers would often use, and even expand, the caverns in the sycamore trees for shelter. It was easier to do this than to make a cabin from scratch. There's even a story about fifteen men on horseback who took shelter from a storm inside a sycamore tree hollow. Alas, these ancient and enormous sycamore trees were cut down in the frenzy of the forestry industry of the eighteenth and nineteenth centuries.

Perhaps the fast-growing and bark-shedding sycamore trees in our part of Ithaca will be left for centuries, until they hollow out and can provide shelter for . . . well, who knows what will be here in the 2320s, 2420s, or 2520s.

The Making of a Green Lake

Any excuse will do to visit Green Lakes State Park, situated some fifty miles northeast of Ithaca. The scenic wonder of these two lakes is celebrated in the name of the park. More accurately, the lake water is the glorious clear blue-green you would expect of the ocean surrounding a tropical island fringed with palm trees. Even in the summer heat of late July, the color seems strangely inappropriate for upstate New York. This local oddity is courtesy of the most astonishing geology, chemistry, and microbiology.

The best place to start is when the glaciers of the last ice age started to melt. At that time, a large glacial river ran roughly west to east near what is now the city of Syracuse. The river raced over a high cliff, probably about four hundred feet high. The force of the water was so intense that it eroded a deep, round plunge pool at its base. And I mean deep. It was some three hundred feet deep. If only we had a time machine to observe this magnificent and enormous waterfall, which was at least twice the height of Niagara Falls. At some point, the river changed direction, abandoning its plunge pool. I have to admit a small complication, that the river deserted not one but two plunge pools. There appears to be no easy explanation for the two pools. The plunge pool at the bottom of the cliff is round, as you'd expect, and is today called Round Lake. The second plunge pool is shaped like a comma and called Green Lake, giving the state park its name.

I'll get back to the green bit in a moment. First, I need to explain how these isolated lakes don't dry up. Sure, they get rain and snowmelt from above, but more than half of their water comes from underground—groundwater that seeps continuously from the bedrock at two levels, about thirty and fifty feet down. This water is very rich in minerals, a concentrated mix of dissolved dolomite, gypsum, and halites. It is a bit like seawater but rich in calcium ions, not sodium ions.

Now for the next step. There are two consequences of having groundwater that's like calcium-rich seawater, and the second bit explains the green. The first bit is that the groundwater is dense—much denser than the fresh water coming in as rain and snowmelt from the top. We end up with two layers of water, one on top of the other, and they never mix. This is most unusual. The water column in most lakes gets mixed up whenever the temperature of the top layer reaches 39.2°F (4°C), the point at which water is at its greatest density and sinks, either increasing to 39.2°F in the spring or declining to 39.2°F in the fall. However, this so-called thermal mixing doesn't work for Round and Green Lakes because the mineral-rich bottom water is denser than the top freshwater, even when the top is at 39.2°F. The result is that, down toward the bottom, there is essentially no oxygen—and no chance of worms, crabs, mussels, or fish. Remember this for later.

Now for the green bit. The reasons why the lakes are an amazing green-blue color like a tropical lake or sea are that the water column isn't clogged up with lots of algae and detritus and, at least in the summer months, the water contains many tiny crystals of calcite, a form of calcium carbonate. These crystals reflect the light very efficiently, and the reflected light is blue-green. So how do these crystals come about?

Remember that the groundwater is calcium-rich. The calcium ions diffuse through much of the water column, which becomes supersaturated with calcium. This means that the slightest chemical tickle will precipitate the calcium. The tickle that counts in these lakes comes from a special kind of bacteria called cyanobacteria. Don't be distracted by the name. Although cyanobacteria means "blue-green bacteria," the blue-green pigments of the bacteria don't give the lakes their color. The bacteria are far too low in abundance for that.

Nevertheless, these bacteria are photosynthetically active. As any biochemist will tell you, a plant or cyanobacterial cell that photosynthesizes releases hydroxyl ions—which is just a fancy way of saying that they make their immediate surroundings a bit more alkaline (or less acidic). And as any inorganic chemist will tell you, alkaline conditions precipitate calcium carbonate, in this case into beautiful crystals of calcite. The calcite forms little concretions around each cyanobacterial cell, distributed throughout the water column. Then each little crystal causes more calcium carbonate to precipitate out from the supersaturated solution.

The main evidence for the calcite crystals is the blue-green color of the water. This chemistry is also visible to the naked eye as whitening of twigs and branches that fall into the lake. More than that, the cyanobacteria and other bacteria can get stuck together, making a reef. This is like a coral reef but called a thrombolite. (What an ugly word, like something out of a medical dictionary!) There's a small shelf of this reef around the edge of the lakes, but the reef extends up to thirty feet into Green Lake and is some thirty feet deep in one place called Deadman's Point.

Slowly but surely, the calcite crystals sink to the sediment at the bottom. Because there are no worms or other beasts disturbing the sediment surface or tunneling into the sediment (I asked you to remember this earlier), each summer's production of calcite forms a coherent layer, sandwiched between the leaf debris that falls from the trees around the lake and accumulates at the bottom during the previous and subsequent fall. The layers of alternating black and white create a perfect record of the conditions in the lake over thousands of years. They are as informative as tree rings. These sediment layers are called varves, a useful word for Scrabble if you ever get two *V*s.

I am sure that everyone who visits Green Lake agrees that it is a stunningly beautiful place. It is made even more special by taking a quick dip into the underlying biology and chemistry to understand how this local wonder works.

August

The calendar tells us that the end of summer is fast approaching, but we avert our gaze. It is simply too hard to imagine the sunhats and tubes of sunscreen on the table by the front door being replaced by woolly hats and thermal gloves; or high-grip, lined boots taking the place of open-toed sandals and bare feet. This month, the cycles of heat and thunderstorms intensify in dripping humidity, while the maximum temperatures gradually decline. In the closing days of the month, the hills may even be shrouded in early morning mist. Meanwhile, the cicadas are buzzing loudly all day, and the crickets are calling day and night. Embrace the cacophony. There is no alternative, for there is no relief from the noise. A second "no alternative" is that the slow, gentle vacation season is drawing to a close. University students are returning to town. The slow trickle during the opening weeks is a prelude to a weekend tsunami of parental cars later in the month. All too soon, students will be swarming in all the usual places, and the academic year will begin again.

By and large, the natural world continues its seasonal cycle unperturbed by the abrupt and arbitrary transitions in human affairs. My essays for August start with a celebration of the creatures that luxuriate in the heat of summer, especially turtles and butterflies

("High Summer"), and I offer no apology for returning to the glory of summer butterflies in the closing essay ("Summer Butterflies"). Between these essays, I take a journey to yesteryear, or more precisely, 390 million years ago, when the local terrain was a shallow sea near the equator and inhabited by unfamiliar animals ("Lamp Shells"). I also take time to consider the amazing biology of the biting insects that plague our August lives ("Blood on the Menu").

High Summer

Some say that high summer is the time around midsummer when the day length is at its peak, but many argue that we need to wait another six weeks or so for "real" high summer. Wait until the bones are warmed right to the marrow, the heart (or many hearts if you are an insect) is warmed to the cockles, and the roots are warmed all the way to the farthest tips. High summer is the time when my bread dough rises in an hour. It is the time when some people start to complain and shove noisy air conditioning units into their windows while others luxuriate in a world that has forgotten what it feels like to be cold.

In the competition for who loves the warmth of high summer the most, first prize must go to the painted turtles. They sit in long rows on rocks or logs just above the water level, a haphazard mix from tiny youngsters to ten-inch-long mature females, their dark shells glinting in the sun. Each turtle extends its long neck, facing its black head streaked with bright yellow stripes to the sun. When space is limited, a smaller turtle heaves itself onto the shell of a larger individual, double-decker style. I've never seen a triple-decker. Perhaps these turtles lack the agility to climb so high, or perhaps the bottom one objects and twitches, toppling the climbers off. Every now and then, a basking turtle plops into the water for a snack. It may swim to the bottom of the pond where it gives chase to a small fish, dragonfly larva, or shrimp, or it may take a munch out of water lilies or water hyacinth plants. At the end of the day, the turtles return to the bottom of the pond for a good night's sleep.

Where there are painted turtles, snapping turtles are also usually found. The snappers remain in the water, where they paddle languidly with all four legs, their long neck outstretched at the front and long tail extended behind. From bow to stern, these beasts can be a good three feet long and can weigh twenty pounds or more. Alas, the snapping turtle is what usually goes into turtle soup (except when cheap cuts of rabbit or chicken are used).

Like painted turtles, snappers are omnivores. Anything from a worm to a small painted turtle or a young muskrat that's looking the wrong way is fair game, together with many aquatic plants. Last weekend, we watched a massive snapping turtle work away at the lily leaves on the beaver pond at Sapsucker Woods. Its formidable jaws clamped onto a leaf, tore away a sizeable chunk, and then, after a quick munch and swallow, the great bulk of turtle repositioned itself for the next mouthful. Then it floated in the water, the top of its carapace exposed and head and tail down, occasionally lifting its head for a breath of air through its snorkel-like nostrils at the tip of its snout.

The slow-motion, drama-free lives of these turtles were accompanied by the regular banjo strum of green frogs. Still, the scene could have ended in drama because some of the frogs were using the lily leaves as floating sunbeds. Let's hope that the snapping turtle's underwater maneuvers create a sufficient disturbance to alert the frogs in good time. All that would be needed is a quick leap to a different lily pad before those fearsome turtle jaws take a bite.

The menu in the water can certainly be varied, but, in high summer, dry land offers one additional item in great abundance: plant nectar delivered as countless tiny meals. Sugar-loving creatures spend their days flying or crawling from one flower to the next, repeatedly extending their tongues to slurp up the heavenly liquid. The nectar comes in receptacles of many colors and shapes. There are the bright yellow wild sunflowers, much more beautiful than the commercial version, the pink joe-pye weed, the purple spotted knapweed, and the brilliant white of wild carrot and snakeroot. Warmed by the sun, beetles, flies, bees, wasps, and butterflies are endlessly busy.

Perhaps the most remarkable nectar-addicted feaster we have seen in recent days is a skipper called the little glassywing. We had superb views of this butterfly while walking through the woodland behind Ithaca College. To the human eye, this site is a bit scrappy, courtesy of electricity infrastructure, including a broad swath of forest cut back to accommodate a line of pylons. The scarlet tanagers playing high in the canopy of hickory trees and the chipmunks racing over the woodland floor weren't bothered by these eyesores—nor were the glassywings foraging in a clump of spotted knapweed just below one of the pylons. This little brown butterfly has spots on its forewing that are totally transparent. When I looked down, the spot appeared to be purple (because the butterfly was on the purple flower of a spotted knapweed) and if I had been able to look up from below the wing (alas, I couldn't do that gymnastic trick), the spot would presumably look as blue as the sky.

It is unusual for butterfly wings to be transparent, but it's certainly not unique to the glassywing. The wings of the glasswing butterfly, a nymphalid and very distant relative of the little glassywing, are entirely transparent, apart from the margins and wing veins. Presumably, the spot on our little glassywing is achieved by the same physics as displayed by the wing of the glasswing. Courtesy of the random arrangement and size of tiny protuberances on the wing surface, light absorption, scattering, and reflection are all minimized. Just for the record, there's no chance you'd come across a glasswing butterfly in New York State. That species lives in Central and South America.

Although they're unusual among butterflies, transparent wings are commonplace in many other insects. Wings are fancy extensions of the insect's exoskeleton, and transparency appears to be relatively easy to achieve. A much tougher task is to make the entire body see-through, as in some marine animals, including some squids and various fish and jellyfish. Generally, being transparent is the very best kind of camouflage, but this doesn't make sense for our little glassywing. I suspect that its transparent spot is a special way to flash its good health (and eHarmony credentials), a variation on the tail of the peacock. It certainly impressed me.

Other butterflies we have encountered recently have been more in-your-face spectacular than the glassywing. Yesterday, we watched a viceroy butterfly feeding from cone flowers in the Cornell Botanic Gardens. This is the butterfly that looks remarkably like a monarch but, unlike the monarch, is not toxic. Any self-respecting predator, such as a blue jay, will steer clear of the harmless viceroy. We also had the treat of watching a feeding giant swallowtail. Its enormous forewings fluttered energetically to maintain its position in the breeze, while its even larger hindwings were perfectly stationary, like a rudder. There certainly is something for everyone in our high summer world.

Lamp Shells

We had been asked to "park efficiently" in the information packet provided before our fossil collecting trip organized by the Paleontological Research Institution. The neat line of a dozen cars on the hard shoulder of the road just north of Moravia demonstrated that we'd all paid attention to the good advice. This was the starting point for our trip. Within a few minutes, we were clambering about the base of a crumbling but mostly sheer cliff of shale, along with about thirty other people of all ages.

It's time to reset your watch by approximately 390 million years— back to the Middle Devonian. Forget Owasco Lake and the other Finger Lakes scoured out by the last ice age, a mere twenty thousand years ago. Instead, float on your back in a warm, shallow sea at the northern end of the Appalachian basin and gaze east toward the towering Acadian mountains. Mountain streams rush down the mountainsides, ultimately reaching a soft, muddy delta where they meet the lapping waves of our inland sea. With your watch reset, your GPS should be telling you that we are close to the equator.

Now flip yourself over. For this bit, make sure your swimming goggles are firmly in place; otherwise, the saltwater will sting your eyes. Look down and you'll see that the seafloor, perhaps a foot or less below the water surface, is smothered in lamp shells. I am confident of this scenario because the surface of the road cutting is made of the compressed remains of that seabed, now crumbling shale and siltstone that can be peeled off by hand. There is no need for a hammer. About one in three of these shale sheets contains fossilized lamp shells. We quickly developed a search image for three forms: *Tropidoleptus carinatus*, which looks like a small cockle; the oyster-like *Athyris* species; and the pretty butterfly shells of *Mucrospirifer mucronatus*. But that is the tip of the iceberg. There are lots and lots of species. Lamp shells take up nearly sixty pages of the *Field Guide to the Devonian Fossils of New York* (Wilson 2014), accounting for nearly one-third of the entire book. If you wanted to be a success story in the Devonian period, your best career choice was to be a lamp shell.

This is a good moment to introduce lamp shells. The meaty bit of the animal is sandwiched between two shells that are connected by a hinge at the hind end. One shell protects the top (dorsal) side of the animal, and the other shell protects the bottom (ventral) side. Apparently, these animals are called lamp shells because some of them look like oil lamps used in ancient Rome. Biologists, most of whom are ignorant of Roman lighting arrangements, gave them a different name: brachiopods. Neither of these names provides a clue to the most important thing you should know about these animals: they don't do much. They just lie on the seafloor (or sometimes burrow shallowly into the sediment) and feed on little bits and pieces in the seawater that waft across their body, a process known as filter feeding. I suspect it is a rather boring way to live, but brachiopods don't get bored because they don't bother with a head or brain.

Brachiopods are easily confused with two other kinds of animals: branchiopods and bivalves. The first is just the result of word confusion. There's little chance to get muddled by the animals themselves, because branchiopods (with an *n*) are fairy shrimps. Although creatures like fairy shrimps had evolved by the Devonian period, they aren't mentioned in our field guide (Wilson 2014) and we certainly didn't see any fossilized fairy shrimp on our trip.

The second opportunity for confusion is much more relevant. Lamp shells are similar in appearance to bivalves, such as cockles and oysters, which also have two shells and no head and also make a living by filter feeding. However, brachiopods and bivalves are very different kinds of animals that just happen to have hit on a similar lifestyle. The main difference is that bivalve shells are positioned on either side of the animal, not on the top and bottom. We uncovered a number of bivalve fossils on the shaly cliff. There are thirty pages dedicated to bivalves in our field guide (Wilson 2014), but the bivalve fossils we found were very small and frustratingly difficult to identify.

History has not treated lamp shells well. Fast-forward from our 390-million-year-old sea to about 250 million years ago. This was the time of the Great Dying, also called the Permian-Triassic extinction event. Scientists argue about why so many species went extinct at this time, but they all agree that lamp shells suffered terribly and never fully recovered. There are about three hundred species today, mostly living in secluded places and at the bottom of the ocean. Today's stars of the headless filter-feeding world are bivalves, with twenty thousand living species.

Blood on the Menu

By the second half of August, the hundreds of acres of old field habitat at the Lindsay-Parsons Biodiversity Preserve are transformed into a mass of brilliant yellow goldenrod, asters, fleabanes, clovers, and milkweeds—all much enjoyed by pollinating insects. This site is definitely worth a visit, but we have learned from experience to go prepared, meaning we need to wear long pants and a long-sleeved shirt, with every inch of exposed skin doused in insect repellant. Butterflies aren't the only insects that enjoy the preserve.

Although the day was forecast to be sunny, ideal for pollinating insects, the conditions changed rapidly to overcast and very muggy. We walked along the path through the brief stretch of woodland from the trailhead to the old fields. As we reached the end of the woodland, we saw a man returning from his visit to the preserve. He looked haggard, as if he'd had a bad night. He jerked his head toward the fields behind him and said, "It's buggy out there."

He was right. Within a minute in the old field habitat, we realized that this was warfare, and we were the enemy. Amid a perpetual whine of mosquitoes and swarms of other unidentified insects, the only strategy was to keep moving. We could easily outpace a mosquito, whose top speed is no more than 1.5 mph, and the blackflies and biting midges are even slower. Still, it wasn't a good day for hanging around birdwatching or identifying pollinating insects.

Keeping moving sounded sensible to start with, but we soon discovered two problems. The first was that we got hot and damp walking in conditions that felt like a steam room in a health spa. Rivulets of sweat were running down our backs, and bug spray–laced sweat seeped into our eyes and dripped off the end of our noses. Unpleasant, yes, but a minor discomfort compared to the second problem: we were advertising our presence to the ravaging hordes. We must have smelled so delicious. All that warm carbon dioxide was now mixed with an enticing bouquet of volatile ammonia, acetic acid, lactic acid, ketones, and more. Another gang of biting insects was ready and waiting as we took each step forward.

Why do they love us so much? Whether it is mosquitoes, biting midges, or blackflies, it is only the female of the species that loves us, and her love is short-lived. She needs just one meal of protein-rich blood to make her eggs. Otherwise, blood is strictly off the menu because the iron in the hemoglobin pigment of blood is toxic and causes her lots of trouble. Most of the time, her food choice matches that of the adult male, which feeds on flower nectar and honeydew. Since they spend so much time moseying around flowers, these insects, which we label as blood feeders, are also pollinators. Not much is known about their pollination services, but a few plant species are absolutely dependent on mosquitoes. One is the blunt-leaf orchid, which likes swampy northern forests, in Canada down into New York State.

I am sure that the legions of insects after our blood at Lindsay-Parsons are the consequence of the large pond and swamp at the far end of the preserve. That's where the blood-fueled females lay their eggs and where the larvae wriggle, feed on microbes, grow . . . and, in due course, join the merry bloodthirsty throng in the old fields. The water habitat at Lindsay-Parsons has nothing to do with the rivers that rush through gorges and tumble down the cliffs as waterfalls near the Finger Lakes—nor is it another of the many bowl-shaped kettle holes, created as the glaciers of the last ice age retreated. Instead, the architect of the pond at Lindsay-Parsons is the American beaver. These astonishing animals dammed a small stream with the branches of trees that they felled with their vicious, bright orange incisor teeth and then dragged into position. Inside the resultant pond, the beavers have built their home, where they are well-protected from foxes and coyotes, which appear not to enjoy swimming for their dinner. At this time of year, the beavers are starting to stock their pond with an underwater larder of tree branches to keep them going through the winter. This is all very well for the beavers, but their engineering kills many trees. We followed the path around the pond, a landscape of utter devastation or haunting beauty, depending on your perspective.

We reminded ourselves that we had come to the preserve for the pollinating butterflies and bumblebees. We did spot a few individuals, but we didn't linger. Somehow, we were far less motivated by these species than by the hawker dragonflies and kingbirds that were gorging on our tormentors.

We returned to the junction between the bug-ridden old fields and the edge woodland by the trailhead. We were about twenty yards from the junction when we saw a young man emerging from the woodland. He was carrying a large camera, with the clear intention of doing some dedicated photography of the beautiful preserve. He called out "How's it going?" with great enthusiasm, presumably wondering why were covered up on such a hot day. Within minutes, he would rue his shorts, sleeveless vest, and sandals.

Summer Butterflies

As we approach the end of August, I realize that it has been a good summer for butterfly spotting. We have clocked up nearly thirty species.

To get some context, let's go back 90 million years. That's when day-flying butterflies are believed to have evolved. They probably arose from something a bit like a silk moth or an owl moth. We tend to talk about "butterflies and moths" as if they are partners of equal standing, like Laurel and Hardy or Abbott and Costello, but butterflies are really a minority. For every species of butterfly, there are seven, possibly eight, moth species. Butterflies get the lime-light because they are fun, and many are very beautiful, whereas most moths are tiny, drab night fliers that are fiendishly difficult to identify.

The butterflies we have been admiring this month make up a story in five parts.

It's logical to consider the skippers first. That's because they are among the most ancient of the butterflies, and they look like it, with their sturdy body and short wings that are often held open when the butterfly is at rest. When I lived in the UK, I tended to think of skippers as the poor relations of the other butterflies. That's partly because of the long-standing arguments about whether they are really butterflies (we now know the answer is yes), but it's mostly because British skippers are a bit limited. Take your pick among the three most common species—the small skipper, the large skipper, and the dingy skipper—all in shades of brown, with the wingspan of the large skipper reaching no more than 1.2 inches (3 cm). Skippers are much more diverse and interesting in the New World. Come to New York State, and enjoy up to twenty-seven skipper species. Some have wonderful names. For example, the Hobomok skipper enjoys the nectar of our backyard cone flowers; the male of the species shows off a bright orange patch on his underwing as he sucks on sweet nectar. My favorite is the silver-spotted skipper, a good two inches (5 cm) in size and sporting a broad splash of brilliant white on the underside of its hindwing. This is special for anyone who started life in the company of small, dull British skippers.

Next come the lycaenids, meaning the blues, coppers, and hairstreaks. My top lycaenid event of this year was a banded hairstreak that settled gracefully on my sunhat for several minutes while I was walking in Watkins Glen State Park in early July. The oak, hickory, and black walnut trees overhanging the gorge at Watkins Glen provide the perfect resource for this woodland butterfly, which lays its eggs on the twigs of these trees. The eggs persist for nine months or so through the winter, unless they are eaten by a chickadee or other hungry beast. The caterpillars emerge in the following spring to munch on the trees' catkins and quickly grow before pupating for next year's butterfly show.

Some lycaenids are much more showy than the predominantly gray-and-brown banded hairstreak. In particular, the spring azure is a wonderful violet-blue. Don't be taken in by the name. Although it is on the wing by early May, this species (or probably cluster of species) is with us through the summer, usually fluttering low in the vegetation beside trails in the forests and meadows. Spring azures are among the many blue butterflies that form a symbiosis with ants. The caterpillars secrete a sugary solution from glands all over their hairy bodies, and the ants lap it up and protect the caterpillars from birds, spiders, and other natural enemies.

Two down, three to go. Let's move on to the nymphalids, also known as the four-footed butterflies. That's because they like to stand on just four of their six legs, curling their forelegs up under their slender body. The four-footeds are very diverse and include the admirals, tortoiseshells, fritillaries, and Baltimore checkerspot (see the essay in "July"), along with the most famous North American butterfly, the monarch. We are getting into the world of big, conspicuous butterflies.

We have especially enjoyed watching one of the four-footeds, the meadow fritillary, which, as the name suggests, flies in meadows and fields, often in large numbers and throughout the summer. The caterpillars eat the leaves of violets, and they would rather die than eat anything else. You'd imagine that Mom would lay her eggs on violets, but she doesn't. She lays them singly and rather haphazardly on twigs, blades of grass, or other plants, and she leaves the tiny caterpillars that hatch to do the hard work of finding a violet plant on their own. Meanwhile, the negligent parents flit around, indulging in the nectar of composite flowers, including dandelions in spring, sunflowers around midsummer, and asters toward the end of the season.

Onward. Our march takes us next to the pierids, meaning the whites, yellows, and sulfurs. Alas, some gloriously named species don't come our way. There's the sleepy orange, which gets as far north as southern Pennsylvania, and the southern dogface, which can be found from Florida to Texas. But we often see clouded sulfurs fluttering in fields and meadows, and we have had excellent views of the orange sulfur, flashing its wings of pale orange with a broad black margin. For better or worse, the most common pierid is the cabbage white, which arrived in the eastern US during the 1860s. It hailed from Western Europe (probably England) and likely arrived in a shipload of turnips or cabbages that berthed in Québec, Canada. The cabbage white promptly spread across North America, probably assisted by railroad transport of fresh vegetables. In just twenty years, it was found coast to coast, except in deserts and the far north. Despite every effort to control this major butterfly pest, the cabbage white continues to flourish. It is by far the most common butterfly in our backyard.

Finally, we have the swallowtails. They are such beautiful butterflies: the eastern tiger, black, spicebush, and eastern giant. The eastern tiger is common, regularly swooping and gliding through the garden on its yellow-and-black-striped wings. Best of all, we have recorded all four swallowtail species in our backyard this year.

September

Inescapably, September is a month of change and transition. Some aspects of September are highly predictable. The rapid decline in day length is inexorable, and we can't help but mourn the disappearance of the many summer visitors, from hummingbirds to ospreys that retreat to warmer climes. Equally, we are cheered by the brilliant yellow of goldenrod flowers and, in the final days of the month, by the bright orange pumpkins that appear for sale at roadside farm stands and in grocery stores. The timing of other shifts is uncertain. We can be confident that the month will begin framed by the vivid deep green of late-summer tree leaves, but we cannot predict how many, or *if* any, of these leaves will start to senesce to yellow, brown, or red before the month is done. It all depends on rainfall, the depredations of insect pests and fungal disease, and whether we have any early frosts.

The first essay for this month of change is devoted to hummingbirds ("Rubythroats"), their presence in our summer backyard, and their September departure. I then consider two insects in turn: the Carolina grasshopper, which stays busy whenever the September sun shines warmly ("The Carolina Grasshopper"), and the elusive harvester butterfly and its meat-eating caterpillars ("The Hunt for the Harvester"). I conclude with a celebration of the quintessential plant of the month: the Canada goldenrod ("Goldenrods"). Somehow, this magnificent plant sums up the closing down of summer and signals the start of the countdown to winter.

Rubythroats

Here in New York State, we have one species of hummingbird, the ruby-throated *Archilochus colubris*. These wonderful birds arrive in mid-May and are gone by the third week of September, back to their overwintering world of Mexico and Central America.

The name ruby-throated comes from a patch of feathers, known as the gorget, at the throat of the male bird. These feathers are dark, almost black, but they contain tiny structures called platelets that are arranged in layers so that they reflect and refract light. If the light hits at just the right angle, it is refracted and appears . . . the most glorious ruby red. This is called structural color because there is no red pigment, nothing like the carotenoids that make male cardinals bright red all over. The male rubythroat with the biggest, most glittering gorget will win every dispute over territory, probably without a fight, and will win the biggest and best female. The females don't have a ruby throat.

Ruby-throated hummingbirds never fail to amaze me. They are tiny, weighing just 0.1–0.2 ounces, the same weight as a lump of sugar, and they have a ridiculously long beak and short tail. Through the summer, they feed on the sugar solution in our hummingbird feeder nailed to a low branch of the red maple tree by our deck. This year, we have been particularly favored by a female. She perches quietly on the base of the feeder, green on her back and head, white with a little gray shading on the front. She dips her fine, black beak into the opening, surrounded by bright red plastic petals, then raises her head. She flies to a dead twig on the maple tree, where she remains perfectly still. About five minutes later, she returns for another snack of sugar solution. Then she is gone, flying swiftly straight over the privet hedge.

In those five minutes of perfectly still birdy meditation between the snacks at our feeder, the hummingbird is processing the sugar. This is tough internal work. She drinks enough to fill the bag-like crop at the front end of her gut. Before she can consume any more sugar solution, the first snack must be delivered, rather slowly, into her narrow, loopy intestines, where it is digested and absorbed across the gut wall into the bloodstream.

By September, the hummingbirds are crazy for sugar. In the week to ten days before they start their long journey south, they need to accumulate lots of fat and increase their body weight by up to 40 percent. There is no time or purpose to forage for insects and spiders during this period when their priority is to fill the tank with fat fuel. Our hummingbirds will soon be gone. Most likely, they will stop to refuel on their journey, especially just before flying across the Gulf of Mexico to their final destination in southern Mexico, Guatemala, or Honduras. A few individuals will sit out the winter in Miami or the Florida Keys. Whatever the details, their eighty wingbeats per second and one thousand heartbeats per minute make these tiny birds the ultimate gas guzzlers.

The hummingbirds are not only one of the migratory wonders of the world but also precision performers. The other day, we were eating lunch on the deck when a rubythroat came to the feeder. We could tell that it was a juvenile, and not our regular female, because it had dark speckling on its throat. It wasn't settling at the feeder but hovered close to one plastic flower, then shifted sideways to the next, and so on, without feeding. Then it flew straight at us and hovered really close, close enough to touch. We could hear the hum of the beating wings: the reason why the first European colonists called them hummingbirds. Pursuing the hum a bit further, the hummingbirds' base wingbeat frequency is about 40 Hz (toward the bottom of the piano range) with readily audible harmonics going up to about 400 Hz (close to the middle A used as the tuning standard at the start of a concert and about the same as the buzz of a female mosquito as she identifies me as her next meal).

Our hovering juvenile hummingbird wasn't just humming. It was also making stabbing thrusts toward us and then backing off, all with perfect control, and it was calling a regular *chip, chip, chip.* We were glad that our eyes were protected by glasses because we sensed that this bird was very cross, out of its brains mad at something. Then there was a slight breeze and the bird returned to the feeder. Still hovering, it drank its fill. Our best explanation is that this young bird was frustrated with the wasps that were also enjoying the sugar solution. Perhaps the wasps left as the breeze rustled the leaves in the maple tree, leaving the feeder free for our adolescent hummingbird.

This is not weird behavior for a hummingbird. There are many reports of hummingbirds coming close to humans, even peering into a human's face. Hummingbirds have excellent vision that is much more acute than human vision. It has even been claimed that they recognize the person who puts out the hummingbird feeder and bother that person if the feeder needs attention. Perhaps the chirps of our juvenile were intended to tell us about the wasps.

Proportionately speaking, hummingbirds have big brains, accounting for some 4 percent of their body weight (compared to 2 percent for humans). Please don't get carried away by this percentage. The total brainpower of a hummingbird is less than ours. After all, hummingbirds can't write about me and then press a button to deliver their scribbling to other hummingbirds anywhere in the world. But then, I cannot fly unaided to Mexico . . .

The Carolina Grasshopper

What a weird butterfly! It flies up, as if from nowhere, several yards ahead of me as I walk along the sidewalk or path, its wings flapping in glossy black with a fetching off-white band along its posterior margin. It flies purposefully in a curve or straight line, then zigzags as if struggling to decide exactly where to land, and finally drops down from a height of ten feet or more to a point farther ahead or to the side in the nearby vegetation. And then it's gone. Sometimes, it is a singleton, but often, especially in a stubble field, approaching human footsteps can disturb clouds of the insects. They look like mourning cloak butterflies in desperate need of remedial flying lessons.

The weird butterfly is, of course, not a butterfly at all. It's one of the band-winged grasshoppers, specifically the Carolina grasshopper (*Dissosteira carolina*). Despite its name, this conspicuous insect is widespread across most of the US and southern Canada.

When at rest, the Carolina grasshopper is nothing special to look at. It is quite big, up to two inches long, and a mottled dull brown-gray all over, providing perfect camouflage on the sidewalk, path, or any bare patch of ground. When the grasshopper prepares to fly, it opens its leathery front wings and unpacks its much larger, paper-thin black hindwings that do all the flapping work. That's how the black-winged insect appears to come from nowhere. As it lands and swiftly repacks its hindwings, it returns to perfect camouflage. It is the classic startle-and-escape response that distracts and confuses any approaching ground predator.

At first sight, it makes little sense that the Carolina grasshopper spends so much of its time hanging around on the sidewalk in harm's way, instead of hiding in the vegetation where it can munch in peace on its standard menu of just about anything green. The explanation is that, unlike many bees, wasps, beetles, and some flies, grasshoppers don't indulge in the trick of shivering their muscles to warm up. For a Carolina grasshopper, getting up in the morning is a slow business, and the daylong progress with its to-do-list depends entirely on the weather. A cool, cloudy, or rainy day will find our Carolina grasshopper a sluggish mover with persistent fog in the brain, while a warm day with bright sunshine sees our grasshopper become an overachiever in its two all-consuming priorities of eating and making more Carolina grasshoppers.

There's much to the task of making more Carolina grasshoppers. Like other grasshoppers, the male Carolinas use sound to advertise their beauty and prowess, but, unlike many grasshoppers, the Carolinas and other band-winged grasshoppers sing with their wings as they fly. The key to their alluring music is in their paper-thin hind wings, which are supported by stiff veins. When specific portions of the wing between two veins are stretched taut, they make a snapping or popping noise. No self-respecting lady could resist the elegant spread of black wings that go snap, crackle, pop. Biologists describe this romantic gesture with the ugly name of crepitation.

Of course, crepitation is just the start. True love can be a long-winded affair for Carolina grasshoppers; mating can last for twelve to sixteen hours. That's it for the male, who presumably then has a relaxing bask in the sunshine and a large meal before he goes back to crepitating. For the female, it is just the start. Her next responsibility is to provision each of some forty fertilized eggs with a large gloop of yolk and some antifreeze and then stick her abdomen in the ground to make a shallow cavity. Lifting her abdomen slightly, she extrudes her eggs, which are stuck together with a maternal glue, into the cavity. She then covers her offspring with soil and smooths everything over with a delicate sweeping motion of her legs. The eggs remain underground and unattended until spring, when the youngsters hatch out and dig their way to the surface. Not all egg batches make it. They run the risk of drowning, being disturbed by a burrowing animal, or getting eaten by a hungry beetle.

And that would be the end of my story, except that crepitation is not the only nomenclatural indignity that our delightful Carolinas have to suffer. Far worse is that they are often referred to as Carolina locusts. This is a grave insult to these good citizens of the natural world. To explain, I need to go back to the beginning of the story. That's some 55 million years ago, relatively soon after the dinosaurs went extinct. This is when the real McCoy grasshoppers evolved, probably in South America. They ate plants in the daytime, had prodigiously long back legs for hopping, and could fly. It was a winning strategy, and these guys—plus their many descendants—conquered the world. Generally, these grasshoppers are called short-horned grasshoppers (meaning they have short antennae) or acridids, and they are related to, but different from, crickets and katydids, which are also known as long-horned grasshoppers. Alas for the rest of us, some acridid grasshoppers can build up enormous populations and then go on a rampage, eating everything in sight, occasionally forming massive swarms. I don't mean dinky little swarms like swarming honey bees but swarms so large (up to billions of insects) that they can darken the sky for hours, even days, as the insects fly through. Acridid species that do this can validly be called locusts: the desert locust of northeastern Africa to southwestern Asia, the red locust of Sub-Saharan Africa, the Australian plague locust, the Central American locust . . . I could go on. These various locusts aren't necessarily closely related to one another, but they are defined by their effect. Some (but not all) short-horned grasshoppers are locusts, just as some (but not all) plants are weeds.

To call the delightful Carolina grasshopper a locust is defamation. Carolinas rarely eat enough to upset humans, and they don't swarm. The Wikipedia entry for this species vaguely alludes to Carolina-inflicted carnage, including "considerable damage" to crops in Saskatchewan in 1933 (I wonder whether the grasshopper was identified correctly) and declares that "there have been no detailed studies on [its] economic importance" (Wikipedia n.d.). Rubbing salt into the wound, several other grasshoppers, including the very common differential grasshopper, are regular and tiresome crop pests but are never called locusts. For the most part, the Carolina is an upstanding and mostly well-behaved short-horned grasshopper that does not deserve to be branded with the name locust.

The Hunt for the Harvester

The harvester is in a class of its own, in a manner of speaking. I will start with the qualifiers. The first is that there is nothing special about its place in the animal kingdom: class Insecta, order Lepidoptera (butterflies and moths), family Lycaenidae (meaning the blues and fellow travelers, see last essay of August). The harvester belongs to the fellow traveler category because it is not blue. It has orange wings with many large, rich brown spots, each ringed in white. It is pretty in its own way but rather small, just an inch across and, by all accounts, easily confused with a skipper. I should point out that I am describing the butterfly from pictures because

I've never seen one in person. That's one reason for the hunt. The other incentive to find this butterfly relates to its behavior, which is unique. I mean unique for North America, though not globally, and that's my second qualifier for the harvester being in a class of its own.

It's not the adult butterfly that makes the harvester special. Like other butterflies, its jaws have been converted into a flexible straw, called a proboscis. When not feeding, butterflies keep their proboscis rolled up below their head, a bit like a hose neatly packed on the side of a fire engine. When the butterfly alights onto a flower, the proboscis is swiftly unrolled, ready to probe deep into the nectary. The harvester adult has a proboscis, just like other butterflies, but it is very short. That seems odd until you discover that harvesters feed on the honeydew released from aphids. Honeydew is the polite word for the poop of aphids and other insects that feed on plant sap. Little honeydew puddles can often be found on the leaves of aphid-infested plants. As the name implies, honeydew is sweet, although it is generally more viscous and less easy to digest than floral nectar. It is an acquired taste. More accurately, it is a taste that has rarely been acquired by butterflies.

The real classiness of the harvester butterfly relates to the caterpillars. The adult female sets up her strange offspring by always laying her eggs singly on the branch of a tree that harbors one particular kind of aphid, a woolly aphid. The wooliness of the aphid is important to the story. These aphids are covered in brilliant white and often curly filaments, making them appear fluffy . . . or woolly. You could say that woolly aphids are the sheep of the insect world, except that aphid wool is made of waxes, which are highly complex lipids, not protein, as in the wool of sheep. Like sheep, woolly aphids flock together. The white, fluffy mat of the aphid flock looks like a fungal growth on the plant, and any insectivorous bird or carnivorous insect that suspects otherwise is educated rapidly by the noxious taste of the wax.

Woolly aphids lead a good life, except when a harvester butterfly lays one of her eggs nearby. The hungry little caterpillar that hatches out lacks the gastronomic sensitivities of most other creatures. It munches its way through the mat of woolly aphids and swallows the soft insides of each aphid, spitting out the wax-covered skeleton. This spitting out is done with great care and perfect aim so the waxy skeleton lands on the caterpillar's back. Then the caterpillar extrudes a short thread of silk from its salivary glands, tying the aphid remains into place. Before long, the back of our munching caterpillar is covered in discarded bits of woolly aphids. The harvester caterpillar is like a wolf in sheep's clothing, but not for long. That's because it takes less than a week for the caterpillar to mature—far faster than other butterfly caterpillars that feed on plant material. But aphids are a superfood, as any ladybug, hoverfly larva, or chickadee knows.

The harvester butterfly has several generations each summer, and both the adults and their wolfish offspring are found all summer long—in principle. The complication is that this butterfly is highly unpredictable and not at all common. In our area, it favors the woolly alder aphid (*Prociphilis tessellatus*). During the summer, we kept our eyes open for the little harvester butterflies and colonies of woolly aphids on alder trees. I am referring here to the speckled alder *Alnus incana,* which predominates in our region both in mosquito-infested swamps, where it grows as impenetrable thickets and, more accessibly, as small trees on the edge of creeks.

Last weekend, we made a concerted effort to find the harvester. After all, the days are shortening, the leaves on the black walnut tree are starting to yellow, and it is often too cold for breakfast on the deck. In short, our days of butterfly hunting are numbered. More importantly for the harvester, the woolly alder aphids will soon be growing wings and abandoning the alder trees, returning to their winter quarters on the silver maple, where they will lay their overwintering eggs.

We traipsed along the section of East Hill Recreation Way that runs alongside Cascadilla Creek. The multiuse trail is broad and asphalted, and it is loved by runners, cyclists, roller skaters, and skateboarders, as well as by families with small children and dogs. At this time of year, the brilliant yellow of the dense stands of goldenrod and the bright purple of the asters under the basswood trees make the trail an especially cheerful place. Our task, though, was to spot the telltale leaf of alder trees that should be growing, here and there, on the creek side of the trail.

We scrambled down from the trail through the dense vegetation toward the creek. Just five yards from the trail, we found a different world. The water ran smoothly over the rocks and stones, two frogs jumped from the bank to the water—*plop, plop*—and a small, orange butterfly darted around. Then it settled, holding its forewings and hindwings at different angles . . . in the usual skipper way. There was no harvester butterfly here, nor was there any sign of woolly aphids on the alder trees.

We tried a second spot. I was briefly entangled in a thorny bramble en route, but it was worth it because the alders at the edge of the creek were particularly fine. What's more, several mats of woolly aphids had taken up home on one of the branches overhanging the creek. Still, there was no sign of any caterpillars. We checked around. No small, orange-brown butterflies, either.

Perhaps we will have no luck this season. We'll try again next year. As everyone says, the harvester is uncommon and unpredictable.

In the meantime, we will look out for the pupal stage of the harvester on branches of alder trees. No one knows for sure how the harvester overwinters, but the pupa is the prime suspect. The pupa is a striking, bright white; is slightly elongated; and has two brownish blobs, as if eyes, above a more irregularly pigmented blob, as if an open mouth saying *ooooh*.

The hunt continues. . . .

Goldenrods

I suspect that, for many people, the word *goldenrod* conjures up an image of one species: the Canada goldenrod (*Solidago canadensis*). This tall plant with myriads of tiny, bright yellow daisylike flowers makes a wonderful show at this time of year, whether in our backyard, along unmown roadsides, or in unmanaged meadows and drainage ditches. It thrives in the full sun, attracting a zoo of different insects—bees, butterflies, beetles, flies, and more—that crawl from one tiny flower to the next and then fly to the next flower cluster, all collecting nectar and pollen.

We can think of *S. canadensis* as the house mouse or house sparrow of the goldenrod world. What I mean is that *S. canadensis* gets along with humans very well. It colonizes any patch of disturbed soil, thanks to its many, tiny windborne seeds, each of which can potentially germinate, take firm root, and spread by its underground rhizomes to form dense, impenetrable monocultures. Individual clumps can persist for a century or more. To keep the goldenrod in our backyard under control, I go on patrol every spring, spade in hand, and dig up any plants that have escaped from the designated goldenrod patch. Some of the escapees have diminutive roots, but many sport strong root systems with long, pink-white rhizomes.

S. canadensis differs from house mice and house sparrows in one important respect: although humans tolerate or resist the mice and sparrows, they tend to appreciate *S. canadensis*. So much so that, a mere twenty-five years after the *Mayflower* brought the Pilgrim Fathers to North America in 1620, *S. canadensis* seeds were transported in the reverse direction to be sold to English gardeners. It appears that nothing much happened, or people didn't notice, until the Victorians decided that every flower border must include a patch of *S. canadensis* goldenrod. By 1849, *S. canadensis* had escaped from gardens and was flourishing in the British countryside. Gardeners across continental Europe and in temperate Australasia were similarly beguiled by the golden beauty, and goldenrod was introduced, for example, to Russia in the 1700s, Germany in 1857, Poland in 1872, Australia in 1935, New Zealand in 1940, China in the 1950s, and Japan in the 1970s.

Everywhere, *S. canadensis* has responded to the hospitality by running amok. There have been various attempts to shut the stable door after the golden horse had bolted into wasteland, native meadows, pastureland, forestry nurseries, and wetland edges. However, the stable door remains open in many countries, where ornamental goldenrod can still be purchased for gardens. Some of the plants available in garden centers and by mail order are said to have been bred to be more showy and less invasive, but I am astonished that people are prepared to take the risk.

One of the chief routes to modify the goldenrod for gardens is to use hybrids between *S. canadensis* and other goldenrod species. This takes me back to the beginning. The common perception of goldenrod as a single species is totally wrong. Come to North America and enjoy the task of discriminating among nearly one hundred species, including twelve that flourish in the meadows, woods, swamps, and roadsides around Ithaca.

Everyone agrees that North America is the center of diversity for goldenrods. There is just one native goldenrod in the UK, *Solidago virgaurea*. We can't call *S. virgaurea* the "British species," though, because it is exceptionally widely distributed throughout Europe, extending into Asia and North Africa. *S. virgaurea* is half the height of *S. canadensis* and bears flowerheads of small yellow flowers. To be honest, it is no beauty; it is as exciting as a ragwort. Even more mortifying, *S. virgaurea* and the invasive *S. canadensis* have ignored their description as different species. They hybridize rather readily, and the hybrid, known as *S.* x *niederederi*, produces viable seeds and has been reported in most European countries, including the UK. So far, it doesn't appear to be highly invasive, partly because it lacks the rhizomes of its North American parent. Perhaps we should be watching that space.

At this point, I want to digress briefly to John Burroughs (1837–1921), a naturalist who wrote persuasively and at length about the environmental and spiritual importance of the natural world. A few years ago, we visited his home, a farm in West Park, in the Catskills. It was late summer, and the meadows were ablaze with goldenrod. Burroughs was an astonishingly influential person. Important people, including President Theodore Roosevelt, paid attention to everything he said and wrote. In particular, he was great friends with industrialists Henry Ford and Harvey Firestone and with the inventor Thomas Edison. This gang of four called themselves the Vagabonds and went on camping trips together, where they thought great thoughts and solved the world's problems.

All of this is relevant to goldenrods. The story starts with Thomas Edison. He was bothered that the supply of rubber from Southeast Asia, which was needed by his pals Ford and Firestone, might be cut off by political uncertainties, and he experimented with various alternative sources of the product. Goldenrod was by far the most promising. Edison bred a twelve-foot-tall high-producing variety. When Ford gave one of his cars to Burroughs, he arranged for the tires to be constructed from goldenrod rubber. Unfortunately, Ford's plans for commercial goldenrod rubber production came to naught, just as his prototype car made of soy-derived plastic instead of steel failed to reach the factory floor.

I can't help but wonder whether the wheels of fashion and need will turn again. One day, goldenrod may be developed as a sustainable source of rubber and other latex-derived products, building on the groundwork of Edison and friends.

October

In this part of the world, everyone knows what October is about: it is the season for fall colors. Our forests are species-rich, and the leaves of the various trees turn different colors in their own time. This means that October offers an ever-shifting patchwork of yellow, orange, red, bronze, and brown. Neither the onset nor the duration of fall colors is predictable, but the order of leaf senescence is fairly reliable, with the black walnut first, and the birches and some oaks among the last. Disconcertingly, the parade of color is especially vivid and long-lasting in town, where the local trees are supplemented by ornamental cultivars and species that have been selected and bred for showy fall leaves and colorful berries. Add in the serried ranks of potted purple, yellow, red, and white "mums" (chrysanthemums) for sale at every grocery store and gas station, and October is a kaleidoscope of color. Meanwhile, the hard business of earning a living in the natural world goes on. Squirrels, chipmunks, and other small mammals are stocking up before the winter begins in earnest; the last butterflies and honeybees of the year forage valiantly for nectar on any day with a shred of warm sunshine; and many birds are on the move to warmer climes.

Two of this month's essays explore the lives of birds. The first examines the strange business of late-season birdsong, focusing mainly on a resident, the Carolina wren ("Autumnal Songsters"), and the third essay considers one of the migrants, the orange-crowned warbler, which passes through at this time of year ("A Relocating Crown"). Trees figure large in the other essays. I consider the black walnut tree, whose fruit is harvested avidly by squirrels ("Black Walnut Bonanza"), and I finish with some thoughts on the ever-changing forest landscape in our region ("In the Carbon Sink").

Autumnal Songsters

We are moving inexorably away from summer and toward winter. The squirrels are endlessly burying black walnuts and other goodies in our lawn, the leaves of our red maple have a reddish tinge, and I have been busy sweeping up box elder leaves from the driveway. The crickets are still with us, but their singing is getting ever fainter. By day, the main sounds are the tooth chattering of squirrels . . . and the song of the Carolina wren.

The full song of the Carolina wren reverberates around our backyard. A wonderful three-note *pidaroo, pidaroo, pidaroo* is repeated very quickly over two or three seconds, then silence, then back to *pidaroo* or interspersed with a frantic buzz. These sounds feel out of place. They belong to the spring, when the birds are disputing nesting territories and impressing their lady loves, before they are overwhelmed by the hard and mostly silent labor of parenting. Now that the young are fledged and parental responsibilities are done, the male Carolina has returned to his blazing song.

One type of explanation belongs to the "can't help it" category. In early spring, long before the temperatures rise, our local birds respond to the rapidly increasing day length by producing lots of testosterone. This potent hormone compels them to sing loudly, fight off intruders, and court females with great fervor.

Birds detect the lengthening days not through their eyes but via a light detector, called the ventromedial hypothalamus, deep in their brain. This tangle of nerve cells contains a pigment called the vertebrate ancient opsin—or VA opsin for short. The opsins are a remarkable family of proteins that change shape depending on whether it is light or dark. The best-known opsins are in the eyes of birds and mammals, including humans, and they are used for seeing. The VA opsin functions differently. It calculates day length from the period in every twenty-four hours that light filters through the very thin bird skull and outer brain tissues to the ventromedial hypothalamus. What's more, this birdy molecular calendar is wired to trigger a cascade of downstream changes in the brain when the time is right, resulting in the overwhelming testosterone surge.

This argument claims that the bird brain calendar has its limitations. It responds to rapid changes in day length, whether it's increasing or decreasing, causing the ventromedial hypothalamus to tell the bird's body that it is time to breed as we pass the late September equinox. The bird starts to get all testosterone-y and starts to sing. Other cues, such as declining temperature, protect the bird from being too stupid, starting to build a nest, for example. (Digressing briefly, bright streetlights and other artificial lights are bad for birds because the extra light confuses the VA opsin in the ventromedial hypothalamus, telling them that the season is more summery than it really is.)

There is increasing evidence that this "can't help it" argument may not be the total explanation. The interspersing of song and the buzz-like calls we hear is usually not the male mixing and matching, but male-and-female duetting. He sings *pidaroo, pidaroo, pidaroo*, and she responds with a *zzzzzzzzzzzzzzzzzzz*. They are telling each other that the kids are off their hands (or wings, perhaps) and it's all still good. They can hang out together through the winter. The duet is also alerting potential interlopers that this backyard and this female are not up for grabs.

It is very possible that some of the singing at the end of the season can be attributed to juvenile males from the last brood. Apparently, Carolina wrens learn their song from their father, possibly starting lessons even before they are born. They begin to sing when they are about two weeks old. This is when they fledge. Carolina wren nestlings develop very quickly, fueled by all those spiders and caterpillars, but their first singing attempts are so pathetic that birders call it "subsong." The youngsters hang around in their parents' territory for up to another four weeks, presumably being very adolescent and annoying, but this behavior does give the youngsters more opportunities to hear Dad sing. Certainly, by the time the young males disperse from their parents' territory, they are fairly good at the real song. It should be possible to distinguish junior from senior, though, because junior is careful to intersperse his *pidaroos* with the occasional *pi-zeet*, which is Carolina wren-ese for "I'm backing off; you're bigger and better than me."

I am pretty sure that all the offspring from the last brood have already dispersed from our backyard. Some of this year's birds may even have bonded with youngsters from other nests, and those new pairs have started wandering around looking for a place to call their own—in other words, somewhere not already occupied by Carolina wrens. Other birds may have sounded or looked good to a recently widowed Carolina of the opposite sex who already has a territory. I guess that, in the Carolina wren world, this is striking it lucky.

The Carolinas are not the only birds with a song to sing in early October. The song sparrows are also in full voice. As with Carolina wrens, the resident song sparrows maintain their territory through the winter, and they are busy marking the boundaries with music. Other autumnal songsters have different things on their minds. The males of the house sparrows and the American robins are bellowing out their superior status to the winter flock, ensuring their preferential access to the next patch of food and their position at the center of the huddle when the first snowstorm comes, as it surely will.

I suspect that much more scientific research will be needed to understand fully why some birds sing in October, but I am sure that the answer is complicated. We should not look for a single explanation for the habits and behavior of wild creatures.

Black Walnut Bonanza

We have enjoyed several days of brilliant sunshine in this second week of October, along with clear nighttime skies. These conditions have driven us to search for our winter blankets and, more importantly, they have brought on the start of the fall colors. Pride of place goes to the black walnut tree, with foliage that has gradually shifted from pale green to lemon yellow as the week has progressed.

Every now and again, a black walnut leaf twirls gently down to the lawn in the crisp air. Then the peaceful, autumnal scene is disturbed by the loud thump of a heavy object falling from high in the tree to the ground. It is one of the tree's many black walnut fruits from this year's bumper crop. Each fruit is about two inches in diameter, the same weight as a small apple, and bright green until it darkens to a dull brown, which is caused by bruising from the thump and general wear and tear on the ground. Other fruits are denied the opportunity to end their days with a big crash followed by decay. That's thanks to another disturbance: an incessant rustling in the tree. The local gray squirrels are chasing around the branches and checking the fruits. When a suitable one is identified, the squirrel clutches it in the forelegs, snips it off the branch with its sharp teeth, and maneuvers the great bulk toward its mouth. Somehow, the squirrel can then dash along the branch and down the trunk to the ground, its enormous booty held securely in its jaws.

Getting to the ground is only the end of Act One. Then comes the big decision: Bury the fruit or eat it? There's plenty of space for burials on the lawn and in the flowerbeds. Our backyard is peppered liberally with squirrel scrabblings and walnut interments. Those buried walnut fruits will come in very handy during the harsh winter to come. Of course, the preferred outcome for the tree is for the squirrels to ignore the carefully concealed fruits, which need a long winter of subterranean damp and cold to germinate. If the gray squirrels left all their buried walnuts (that won't happen!), our backyard would be jungle of black walnut seedlings next spring.

Far worse for the tree is the alternative and frequent choice of the squirrel. All too often, the squirrel succumbs to temptation and consumes the walnut, here and now. Hungry or imprudent, the squirrel rushes to an open spot, where it can keep an eagle eye out for the approach of a greedy friend or relation. Then it swiftly tears the shell apart with its teeth. This noisy business, a bit like the loud crinkling of aluminum foil being scrunched up, can be heard again and again through the day. The hard, brown nut at the center is a substantial and energy-rich snack, and it is hastily consumed. Satisfied, the squirrel scampers off to climb back up the walnut tree. All that is left is a small pile of bright green walnut shell chippings. Over the following days, these fragments will turn brown and soggy and ooze a purplish liquid. Our backyard is littered with these neat little piles on the lawn, the deck, and the driveway. And there are many more mounds of discarded black walnut shells here and there at various stages of decay on the roads and sidewalks in our quiet neighborhood.

It's worthwhile to spare a moment's thought for the jaws of the gray squirrel at this time of year. The hinge-like joint connecting its lower jaw to its skull must be stretched to its limit every time the squirrel carries the large and heavy fruit. What's more, the force applied through the squirrel's jaws and teeth in order to break open the walnuts must be immense. Humans need a hammer to get into them. An alternative strategy for harvesting black walnuts is to leave the fruit in a bucket until they get mushy, at which point the shells just slip off. If you try this, be sure to wear strong, water-proof gloves; otherwise, your hands will be stained a deep blue-brown. Meanwhile, the squirrels have sore jaws, discolored teeth and paws, and a good stock of food for the winter.

Let us return to the black walnut tree because it is not only the gray squirrels that are causing a rustling in the foliage. The red squirrel is also very much in town. Earlier in the week, we watched several reds chase one another up, down, and around, all at great speed. Most likely, they were youngsters preoccupied with the tasks of playing and mock fighting. Unlike the gray squirrels, who continue the complexities of social interactions throughout their lives, this activity is short-lived for the reds. We haven't seen them in the last couple of days, and I suspect that the youngsters have gone their separate ways to stake out their individual territories. All in all, most adult red squirrels lead lives of near-monastic isola-tion in fiercely defended territories. There are exceptions, though. A female allows males into her territory for a single day during early spring and again at midsummer, before shooing them away so she can raise her two litters alone. And in some rich habitats with plenty of food, the territories become porous and overlapping, although individual squirrels continue to avoid direct contact. Our observations over the years suggest that the conditions in our backyard favor strict territories.

The female red squirrel that owns our backyard is very busy. Her preoccupation is black walnuts, not her offspring. She careers around the tree, then nips off a suitable fruit, letting it fall to the ground. Then she races down and manhandles the great fruit back to her larder, located in the narrow strip between our garage and the garden fence.

When we checked the red squirrel larder earlier today, there were more than a hundred black walnut fruits, together with some spruce cones. The squirrel must have spotted us because she came racing down the spruce tree, chattering fiercely. We backed away to demonstrate that we had no designs on her hoard. Life is neither easy nor placid for the red squirrel, but if she can protect her store, then she will be set up for the winter.

A Relocating Crown

I am not writing about a worrying dental problem but a much more interesting and decidedly mobile crown in Allan H. Treman State Marine Park at the southern end of Cayuga Lake.

It was early on Sunday morning, and we had the trail almost to ourselves. As we swished our way over a carpet of yellow, brown, and red leaves in brilliant October sunshine, we passed solitary song sparrows singing out the boundary of their territories every hundred yards or so, and we heard the loud caroling of the occasional Carolina wren. It was calm at the lakeshore, and the narrow concrete spit extending from Lighthouse Point to the Cayuga Inlet

Lighthouse bore a long line of birds: many double-crested cormorants, together with ring-billed, herring, and black-backed gulls. The gulls will stay with us all winter, but we anticipate that the cormorants will be departing for warmer climes before long.

But then there was a surprise. Near one of the song sparrows, another bird was flitting in the bushes. It was definitely small and decidedly nondescript, apart from a bright yellow patch under its tail. Its breast was perhaps a little streaky (it was difficult to be sure), and its beak was narrow and pointed. It was definitely a warbler, presumably taking a break on its journey south. A quick check through our bird book showed that only one species sports a bright yellow bottom: the orange-crowned warbler or, more colloquially, the orange-crown.

A little more detective work was needed to work out where our orange-crown had come from and where it was going. The summertime distribution of the orange-crown is like an upside-down *L*, extending across the midlatitudes of Canada and Alaska and down the west coast of North America, inland as far as the Rocky Mountains. It's a bit more complicated than that because there are four subspecies of orange-crowns. Our bird was almost certainly subspecies *celata*, the most drab of the orange-crowns. Our little *celata* could have flown straight down from northern Québec or it could have come across from Alaska on its way to Florida, the Mississippi Delta in Louisiana, or Mexico.

Orange-crowns almost always travel by night and are usually alone, but they will occasionally travel in small groups. As with the many other birds that migrate under cover of darkness, orange-crowns navigate using the stars. More specifically, they use the North Star: the one star that is stationary in the sky all night. The constellation of Orion is also useful because its apparent position shifts little as Earth rotates. Stargazing is a great way to navigate on a clear night, but it's not so smart when it is cloudy. Nevertheless, our little birds can keep going, thanks to a second navigational system that is weather-independent; they can make use of Earth's magnetic field. I am pretty sure this hasn't been studied in orange-crowns, but other warblers, notably the European reed warbler, have tiny, iron-rich crystals of magnetite close to their nares (you could say in the nose). Any change in intensity and possibly direction of the magnetic field is communicated from these crystals to sensory nerve endings associated with the nares and beak, and the nerve impulses are transferred, in an instant, to the brain. Presto! The bird has information on where it is and where it is going.

So much darned cleverness has to be learned by the youngsters while sitting in the nest at the bottom of a scrubby bush in the forest—and during the short period between fledging and the four thousand–mile journey south: how to fly; how to find all those scrumptious beetles, caterpillars, and spiders; how to avoid the local hawks and any cats in the neighborhood; and how to recognize and, for the boys, reproduce the sweet, trilling song. On top of all that, they also have to learn how to read the stars. I suspect that interpreting the electromagnetic field generated as Earth's molten iron core swirls around is hardwired in their little warbler brains. It's a crash course for survival during their first two or three months of life.

There's one more important issue to address about the little orange-crown that was hopping about in the bush last Sunday morning: Why is this ultimate LBJ (an honorary little brown job that is in fact gray) called an orange-crown? Almost none of the pictures and photos of the bird display brilliant orange headgear. Exceptionally, the entry for this species on the Cornell Laboratory of Ornithology (n.d.) website (All about Birds) has a photo of a male turning its head to the camera and displaying a diminutive pale orange patch on the top of its head. The accompanying text informs us that the orange patch can become evident when the bird is excited or agitated, but it is not usually visible.

I strongly suspect that the bird was named from a dead specimen, either shot or trapped, allowing detailed inspection of the diminutive orange patch. The authority was Thomas Say (1787–1834), whom I have encountered before as the much-lauded father of American entomology. Say was an intrepid explorer who identified and named the orange-crown while on an official expedition to the Rocky Mountains in 1819–20. He was also something of an idealist. In 1826, he was one of several hundred who joined the New Harmony settlement in Illinois with the dream of jump-starting a new moral world. Of course, the settlement collapsed in quarreling unhappiness by the end of the decade, but Say didn't get involved with the inharmonious infighting. Instead, he focused on his natural history studies, including preparing his groundbreaking three-volume treatise on American entomology. He stuck around in New Harmony, even though the utopian settlement had dissolved, pursuing his natural history studies, until he succumbed during an outbreak of typhoid fever in 1834.

Meanwhile, the inappropriately named orange-crown today flourishes with the conservation status of least concern. There is nothing better for the orange-crowns than all that shrubby under-story that grows up after logging and is then left alone (thanks to the no-fire management policies of the last century).

Nevertheless, there is no place for complacency when it comes to the orange-crowns. There are many ongoing changes in the land-scapes where these birds breed. Forest management practices are shifting to favor elimination of the highly flammable bushy under-growth where they nest, and the increasing incidence of intense wildfires and heat waves in the west are taking their toll. The progno-sis for very few species can be assured in our rapidly changing world.

In the Carbon Sink

The area around Ringwood Ponds is a pleasure to visit in the fall. The trails weave through mature maple, birch, oak, and hemlock trees, with saplings of all sorts in the understory. Dead tree snags are riddled with woodpecker and insect holes, and windthrown trunks lie untouched on the forest floor, supporting a world of wood-decay fungi of various colors and shapes. Newly fallen leaves litter the ground, chipmunks and squirrels scamper around, and blue jays call from the upper canopy. It is a magical place with a timeless feel to it.

But this landscape is far from timeless. It is obvious from the terrain that it hasn't always been like this. The trail takes us along eskers, ridges of now well-stabilized sediment that was deposited by the Wisconsin ice sheet as it receded between ten and twelve thousand years ago. The dips on either side of the eskers form temporary ponds that usually dry up in the summer and are reformed with snowmelt each spring. Another unambiguous sign of the past ice ages are the several kettle holes: round depressions scoured out by a block of glacier ice that became covered in sedi-ment and then melted. These kettle holes gave the place its name: Ringwood Ponds. Amphibians love the site for its ponds and swamps, and the Big Amphibian Courtship Concert takes place at Ringwood Ponds every spring (see April, "Spring Peepers").

As the ice sheet and glaciers receded, the barren land was colo-nized by plants and animals, ultimately leading to forests. I wish

I could add that our old-growth forest is the direct descendant of those first postglacial forests, but there is a complication. That complication crossed the Bering Strait from southern Siberia to North America between eleven and twelve thousand years ago, and it then reproduced and dispersed, eventually reaching what is now New York State about nine thousand years ago. I am writing about humans.

These humans gave rise to the Indigenous peoples of North and South America. In our region, they engineered the landscape with fire. Regular, purposeful burning of the forests killed fire-sensitive trees and the understory, leaving a park of magnificent trees with thick, fire-resistant bark. These trees were mostly nut-producing red oak, shagbark hickory, and sweet chestnut, along with pine and hemlock. There was no need to bushwack your way through undergrowth when hunting for deer, and the groves of nut trees supplemented the corn and bean crops grown in small clearings. It appears that this type of landscape engineering was sustainable and supported relatively large human populations. In the first written records kept by Europeans in North America, the land in this area was occupied by the Gayogohó:no̱ʼ (pronounced Guy-yo-KO-no), which means "from the swampy land" (Jordan 2022).

There are a few very steep and inaccessible places in the local region where the mix of tree species is best explained by human engineering that predates European settlement, but Ringwood Ponds is certainly not one of them. To understand Ringwood Ponds, we must consider the complications of human history. European colonists did not settle in our patch of New York State until after the American Revolutionary War (1775–1783). The local Indigenous people had backed the wrong side (the Brits) during the war. As retribution, they were driven out, and the land was given, in ten-acre packets, to veteran American soldiers. In the eyes of most Americans, Indigenous people had no legal ownership of the land because they did not enclose it. The white settlers chopped down the forests to farm crops and establish sheep pastures. These ventures were ultimately unsustainable because the soil, scoured by glaciers, was poor and the agricultural methods were primitive. Within a few generations, residents started to move to the western frontier, meaning the rich river basins of Ohio, and then even farther west. The abandoned farmland in New York gradually reverted to forest. Fragments of stone walls, which had been built to enclose sheep, and the foundations of cottages are evident all over the place in the forest.

The forest at Ringwood Ponds is designated as old growth because it has not been logged since the 1860s. Much of the forest in this area is more recent than that, including trees that have grown back over farms that were abandoned well into the twentieth century.

This brings me to the carbon sink of my title. The forests of New York State and New England represent an important sink for the carbon dioxide in the air. As the trees grow back over old agricultural land, they suck in carbon dioxide, trapping the carbon in wood that is laid down as they grow. However, as we see in the old-growth forest at Ringwood Ponds, trees eventually die and their wood is consumed by fungi and insects, converting much of that carbon back to carbon dioxide. There is excellent evidence that the amount of carbon dioxide New York State forests are gobbling up (in other words, the strength of the carbon sink) is declining because the forests are maturing with increasing numbers of dead and decaying trees.

What this means is that New York State is a strong carbon sink not because it has lots of forest but because much of that forest is young and actively growing. What's more, our forests are merely recouping carbon lost from the land by deforestation over the last few centuries. It is sobering to think how the consequences of how we manage the land takes effect over very long periods of time—often longer than we can readily comprehend.

November

It is all too easy to decide that November is grim. This is the time of year when the warmth and much of the color are stamped out of our world. The days are shortening, and we are at the mercy of northerly winds and cold rain, perhaps mixed with wet snow, and of gray frosts and fog. The fainthearted have retreated. Many birds and the monarch butterflies have migrated south; other creatures, including the chipmunks and many insects, are lying low in a secluded spot, often in a dormant state. For the rest of us, it is time to start the business of embracing the winter. We keep on going outside to see and hear what is still happening, and we pay attention to the skies. On some November days, the clouds on the horizon are drenched in the most splendid pink and red, yellow and purple, marking the sun's daily arrival in the east or departure in the west.

The essays for November are a celebration of the wonders that remain after summer has definitively ended. The first essay concerns the Canada geese that live in our region, other parts of the US, and, thanks to the whims of a king, in Britain ("Wild Geese"). Then I consider the remarkable flowers of one of the very few late-flowering plants in our area ("Witch Hazel"). The lives of the cardinals in our backyard are a story of endless intrigue, and I share the story of the time when the male who owns our backyard replaced his partner with a new, more docile mate ("All Change"). The last essay explores the diverse perceptions of one plant, a mullein: a valued medicinal plant, an invasive, and a food resource that supports weevils and woodpeckers in search of a weevil snack ("The Greatness of the Great Mullein").

Wild Geese

The grounds of most English stately homes have been landscaped to include a large lake. Many of these lakes are now a refuge for wildlife and great places for birdwatching. Everything is worth a mention except for one ubiquitous bird, the Canada goose. There are always rafts of Canadas, large birds with long black necks, a broad white stripe running under the chin from cheek to cheek, white fronts merging into light brown on the belly, and brown wings. "Handsome birds, really, but as common as muck and they shouldn't be there anyway," we would say.

Please blame King Charles II. Some specimens from the Atlantic seaboard (that's important for later in the story) were transported to his burgeoning wildfowl collection in London's St. James's Park. I wouldn't fancy the ten-week transatlantic crossing while caring for a gaggle of endlessly hungry, endlessly pooping, and endlessly ill-tempered Canadas cooped up in a pen on the deck of a seventeenth century sailing ship. I am sure that Charles was pleased because the newly arrived Canadas let loose in St. James's Park were bigger than any native goose and, as I've already mentioned, decidedly handsome. Before long, everyone who was anyone, meaning the owner of a country estate, had to have some Canadas, too. Over the years, the Canadas prospered and went down-market, as it were, to every available habitat that combines standing water and nearby grassland—including the manicured lawns of city parks, university campuses, golf courses, and so on.

But Canadas are like gray squirrels. They are much, much more interesting on their home turf of North America than in their adopted land of the UK. At one time or another during the year, almost everyone on this continent shares their life with Canadas. Go to Canada, and the Canadas are summer visitors. Go to Texas, Louisiana, or New Mexico, and they are winter visitors. Here in New York State, we have the pleasure of their company all year round.

One of this week's encounters with Canada geese was during an early morning walk along the eastern shore of Cayuga Lake. The mist was still clearing from the steep hillsides, blanketed in forests of naked trunks and branches; a storm earlier in the week had ripped the few remaining leaves from the trees. The lake was a placid, gray millpond. Close to the shoreline, a raft of Canadas was drifting, apparently aimlessly, with the occasional companionable honk. We could see additional rafts up and down the lake, along with many ring-billed gulls and a few mergansers. Another goose meeting was at Sapsucker Woods at the northern edge of Ithaca. The woodland there includes a large pond; the stumps of recently felled trees chewed to a sharp point remind us that it is an active beaver pond. As we approached, a pair of wood ducks took off explosively from the water. The other birds, though, were not perturbed. The red-winged blackbirds continued to call from the branches of trees drowned in the beaver-engineered flood, and two separate rafts of gently honking Canadas glided by.

At this time of year, the local population of Canada geese balloons. The birds that breed locally stay put, and they are joined by large numbers that spend the summer farther north. Some of these migratory birds stay with us through the winter. Others are just passing through, always flying south. A skein of Canadas could be an extended family commuting a few miles between sleeping quarters on the water and daytime feeding in agricultural fields or at the local golf course, or it could be a group from Baffin Island on a journey of thousands of miles southward to the Everglades of Florida or the swamps of Mississippi.

Canada geese on their home turf in North America lead more interesting and wilder lives than they do in the UK, and they are also much more diverse. There are a total of seven recognized subspecies, differing mainly in size. The subspecies in the UK (*canadensis*) is the same as in Ithaca, and it can weigh up to 11 lb (5 kg). Two other subspecies are certainly worth a mention. There's the giant, known formally as subspecies *maxima*, which weighs up to 22 lb (10 kg). *Maxima* is very special because it was declared extinct in the 1950s but was then rediscovered a decade later in Minnesota. (The traders who supplied King Charles II wouldn't have known about *maxima*.) Then there's the lesser Canada goose (*parvipes*), maxing out at just 6 lb (2.7 kg), which breeds in western Canada and overwinters in Oregon, specifically in the Willamette Valley, just south of Portland. If I had written this piece at any time before 2004, I would have added an even smaller Canada goose that breeds in the far north, a Canada that is barely larger than a mallard duck—but the taxonomists have been at work and decided that these duck-sized Canadas are a different species altogether, the cackling goose.

Just imagine if King Charles II had acquired cackling geese instead of *canadensis*. Feeding the ducks in the UK would have been a much less exciting experience. The mallard-size geese might hiss and bite, but somehow a 2–4 lb goose doesn't have the oomph as a 11 lb goose. Charles would probably have gone for *maxima* if he had the choice. That would have been like feeding pelicans.

Witch Hazel

It was John Keats (1820) who claimed that autumn is the time for "budding more, and still more, / later flowers for the bees." But not indefinitely. As October progresses, the goldenrod flowers turn to gray seed capsules, the brilliant yellow petals of the black-eyed Susans shrivel and fall, and even the purple flowers of the New York aster fade out. The small white flowers of calico aster and daisy fleabane make it to the opening days of November, but then time is up for them, too.

But there is one magnificent end-of-year bloomer: the American witch hazel (*Hamamelis virginiana*). Don't be taken in by the *virginiana*. American witch hazel is an east-coaster with a far greater range than the state of Virginia. It extends from Nova Scotia in Canada down to the Gulf of Mexico. For most of the year, our witch hazel is the underdog of the tree world. It cowers in the understory beneath the big, serious trees, such as maples and beeches, its several thin trunks and even thinner crooked branches spreading out, as if the tree were in perpetual danger of losing its balance. The summer leaves are unremarkable. The dull green ovals with lightly indented margins are nothing to write home about, except that they are a bit like the leaves of hazel trees found in Britain—so much so that the first English colonists decided this new tree was a hazel tree, like the common European hazel *Corylus avellana*. They were hopelessly wrong: the American witch hazel (of the family Hamamelidaceae) is a very different beast from the hazels, which are birches (family Betulaceae).

In early November, the American witch hazel comes into its own. As everything else is closing down for the winter, the witch hazel starts to flower. The four petals of each flower are a wonderful buttercup yellow, but they are thin and twisted. On a sunny autumnal day, the petals extend, sticking out like spikes, and the flowers become fragrant—unmistakably the same smell as witch hazel water in the medicine cabinet (more on that later). It is all amazingly exciting for small pollinating moths, as well as for winter gnats and possibly the last of the bees. Then, when it's cold and frosty, the petals curl up, as if battening down the hatches at a time when pollinators are too sluggish to function. Come another dash of warm sunshine and the petals unfurl to advertise their wares to pollinators.

It is not only the flowers of the witch hazel that refuse to be in any seasonal hurry. The fruits play it long, too. By the end of the year, the flowers will be done, each replaced by a small capsule containing either one or two seeds. Initially green, the capsule gradually turns brown and woody and remains attached to the branch for a full year. This means that last year's seeds are being dispersed at the same time as this year's flowers are on show. Seed dispersal, witch hazel style, is best described as blast-off. The woody capsule snaps open and the seed is discharged with great force, sometimes up to thirty feet away. Then the seeds wait for at least another year before germinating. The witch hazel is in no hurry.

It is most unusual to see both flowers and fruits on a tree at the same time. The great Swedish taxonomist Carl Linnaeus was very impressed by this trick, and that's why he coined the name *Hamamelis*, meaning "together" and "fruit." I have not discovered whether Linnaeus made his formal description of *Hamamelis virginiana* during one of his visits to North America or if he depended on specimens collected for him during one of the many expeditions of his students. Either way, finding these small trees bearing flowers and fruits simultaneously must have been very special. There are no native witch hazels in Europe.

Let's now turn our attention to the witch of witch hazel. The consensus view is that this has nothing to do with witches. The "witch" is a derived spelling that started with *wiche*, a Middle English term for "bendy" or "flexible." By most accounts, witch hazel started off as wych hazel, as also used for wych elm, the only bona fide native elm in Britain. The early American colonists must have been delighted to find a hazel-like tree with super-bendy branches that could be twisted, bent, and split without breaking—and could be used for wattle, fencing, and basketmaking. By chance, the spelling gravitated toward witch hazel, while wich hazel, which hazel, and wych hazel slowly fell out of use.

The Indigenous people of North America knew something else about witch hazel: its leaves and bark could be used to alleviate skin inflammations, from bruises to insect bites and poison ivy rash. Once they learned this, European settlers enthusiastically added witch hazel to their pharmacopeia. Then, in the nineteenth century, a missionary by the name of Dr. Charles Hawes discovered that he could make a fragrant extract by collecting the steam released from a boiling pan of witch hazel twigs. Before long, Hawes's extract became a commercial enterprise, and the rest is history . . . except that it sold better with a more generic label: witch hazel extract.

I have vivid childhood memories of the near-miraculous power of witch hazel to eliminate the pain of every bump and bruise. Disconcertingly, there is no persuasive scientific evidence that witch hazel extract has any efficacy at all.

All Change

By the third week of November, winter had definitively arrived. That was when we had our first winter weather advisory, informing us of several inches of snow, followed by freezing rain, in the coming twenty-four hours. Woolly hats, scarves, and gloves are now an essential part of our outdoors wear, and we have set up the bird feeder in the backyard. The birds will be glad for the extra calories in the nuts and seeds in the bird food to survive the increasingly long and cold nights.

Within hours of attaching the feeder to the nail on the branch of the maple tree, the usual suspects arrived. The alpha male gray squirrel was the first visitor, but he failed to get past the outer wire cage to the cylinder of food. Then a junco settled in beside one of the holes in the cylinder and munched stolidly for several minutes before being disturbed by a party of chickadees and tufted titmice. Our local pair of cardinals then came to check out what was going on. The bright red male perched on the maple tree for a minute or so, then flew down to the ground under the feeder in search of any pickings. Perhaps he sized up the dimensions of the wire cage and realized that he was too big to squeeze through to the food—or perhaps he remembered from last year that his best bet is to feed on the waste below before the squirrels arrive and hoover it up.

The female cardinal was in attendance, too, perched primly on the branch of the maple tree. This was exactly as one would expect because cardinals remain paired up through the year. Last spring, our female, being smaller than her partner, discovered that she could get through the wire cage to the food. Would she remember? Before I go any further, I need to explain that our female cardinal has been quite a character. This became very apparent last spring when the woodpeckers were starting their courtship drumming. We noticed that some of the drum routines didn't match any of the rhythms of our local woodpecker species and sounded strangely tinny, almost tinkly. Initially, we localized the sound to the backyard of our southerly neighbor, but then it started up just over the privet hedge along the west side of our backyard. It wasn't a woodpecker. It was our female cardinal. She was attacking her reflection in the side window of our west neighbor's garage. She would jump up from the window ledge, flapping her wings and thrusting her head forward until her stubby beak knocked into the glass. Her feet would briefly land back on the window ledge, and then she repeated the attack, again and again, for a several minutes. It must have been exhausting. We started to call this activity her Daily Hate, except that sometimes she was driven to a fury by the interloper in the garage window several times a day. Occasionally, her mate would join her at the window ledge, but his reflection was far less aggressive, and he usually left her to deal with her adversary on her own.

I think this is telling us that cardinals aren't terribly bright. Our incensed female lacked the brainpower to recognize herself. I am reminded of the mirror test used by scientists who study animal behavior. Mark the animal's forehead with a red spot and then give the animal a mirror. If the animal pokes at the red mark, then it is deemed to be sufficiently self-aware to understand that the image is itself. Chimpanzees, elephants, dolphins, and magpies pass the red spot test, but dogs and octopuses fail. Babies are in the dog-octopus camp; most young children develop the capacity to recognize their reflection (and a red spot on their forehead) by their second birthday.

You'd imagine that all the time and energy that our female cardinal spent on her Daily Hate through the spring and summer would detract from her commitment to incubating her eggs and tending her young. You'd be wrong. Despite her obsession with the bird in the window, our cardinals had two clutches, and they successfully raised several offspring.

The incessant pecking at the window must have rattled her brains. Every time she slammed into the window, the front of her brain would be compressed, and the back would be stretched out. I wondered vaguely if the skull of a cardinal might have some spongy bone to absorb the shock, as has been reported for woodpecker skulls. Whatever way you look at it, the behavior of our female cardinal can't have been doing her much good.

Back to the female eyeing the bird feeder from the branch of the maple tree. She looked different. She was perkier, more upright, and more attentive to what the male was doing than our familiar female. Goodness! The only explanation is that we have "all change" in our local cardinal world. Our male has acquired a new mate. Perhaps the previous female had been ousted because she failed to pay attention to her real adversaries, or she may have succumbed, overtaken by the combined rigors of parenthood and the Daily Hate.

The new Mrs. Cardinal dutifully followed the male to the ground to peck at the discards from the feeder. As we have watched the new pairing, we have observed that she trails her partner most decorously. She does not appear to have much of a mind of her own. Unlike her predecessor, she is unlikely to try squeezing through the squirrel-proofing mesh to access the food in the feeder anytime soon. Let's hope she also doesn't become obsessed with windows.

Getting back to the feeder, we noticed that the food level in the cylinder was declining precipitously. We sighed heavily and blamed the usual suspects: a gang of house sparrows. But we didn't see any sparrows. It was very strange. A day later, the big guzzler was caught in the act. It was this year's red squirrel, slipping easily through the mesh to enjoy a leisurely meal. Red squirrels are with us always, but this is only the second bird-feeder-savvy individual we have encountered (the first was in 2012).

Here's a problem! Our red squirrel won't be hibernating. I look at the sack of bird food in the basement and mentally rename it "the sack of red squirrel food."

The Greatness of the Great Mullein

According to Richard Mabey's *Flora Britannica* (Mabey 1996), the great mullein has many names. Aaron's rod, hagtapers, candle-wick plant, and Our Lady's candle remind us of the way its stalks used to be dipped in suet and then lit to make firebrands. Other traditional British names are Adam's flannel and Our Lady's flannel. As Mabey (1996, 329) comments discreetly, "Its leaves are used accordingly."

I am still a long way from finishing with names. Great mullein, a native plant of Britain and temperate Eurasia, is also known as common mullein and has the Linnean name *Verbascum thapsus*. Acknowledging the big and decidedly hairy gray-green leaves, *Verbascum* comes from the Latin *barbascum* ("beard"), and mullein comes from the Latin *mollis* ("soft"). If you are curious about the *thapsus*, go to your atlas. You will find the tiny island of Thapsos off the coast of Sicily and renowned in certain circles for its distinctive Bronze Age artifacts of the Thapsos culture. So, when Linnaeus made the first formal description of great mullein in 1753, was he on holiday on Thapsos? No, the story is more ancient than that. It seems Linnaeus knew that an ancient Greek philosopher called Theophrastus (371–287 BCE) described the plant during a visit to Thapsos.

Altogether, the great mullein is anything but an any-old-plant. It was an essential ingredient in the repertoire of every herbalist from the most ancient times. Go anywhere and anytime in Europe over the last two thousand years or so, and mullein was used. Even today, in our world of marvelous modern medicine, from MRIs to monoclonal antibodies and mRNA vaccines, great mullein keeps on going. Health stores continue to sell bottles of dried leaves or flowers, often pounded to a fine powder, and easy-to-swallow capsules of mullein essential oils. The more restrained advocates recommend a cup of mullein tea to alleviate respiratory disorders, including a chesty cough and bronchitis, but be sure to filter the tea before drinking; otherwise, its beardy hairs will scratch your throat. A more expansive view is that mullein is anti-inflammatory, antioxidant, anticancer, antimicrobial, antiviral, antihepatotoxic, and anti-hyperlipidemic, meaning that this plant can cure just about everything. No one knows the basis of this medicinal wonder plant. Mullein is packed full of various glycosides, flavonoids, and saponins that are highly toxic, designed to deter marauding munchers. In my opinion, caution is recommended.

One fact for sure is that the early European settlers in North America brought great mullein seeds with them for their medicine cupboard. Colloquial American names tell us about other uses: the flannel leaf plant for the restroom and Quaker's rouge for the young women of this religious group who rubbed the leaf into their cheeks, causing a fetching maidenly blush of mild dermatitis. There are also historical records that show the leaves were used to pad shoes and clothes for extra insulation during the bitter North American winters.

The good news–bad news about great mullein is that it produces lots of seeds, often more than one hundred thousand per parental plant. The seeds are remarkably hardy, germinate freely in response to light and moisture, and can stay viable in the soil for more than a century. This means that the multipurpose savior of the early colonists is now a pernicious weed across much of the

US, exploiting many habitats, from pasture fields to urban brown sites and backyards.

Except that great mullein is welcome in our backyard. It is the classic biennial plant (meaning that it lives for two years). We love its rosette of large, gray-green leaves in the first year and the tall spike of sulfur yellow flowers in the second year. At flowering, each elegant plant is at least four feet tall, and some plants reach six feet. We also enjoy the many pollinating insects, including bumblebees, halictid bees, and hoverflies, that visit the flowers.

So far, so good. But this year, our wonderful show of self-seeded mulleins has been infested with the mullein weevil (*Gymnetron tetrum*). We first spotted the adult weevils in June: lots of dull-brown beetles with the tell-tale long weevil snout, scrabbling around on the flower spikes. Once a flower is pollinated, the female weevil deposits a few eggs into the developing seedpod, and the larvae stay there, consuming the seed nutrients, until next summer. They will also eat one another if the mullein rations get low. Siblicide is fair game in the mullein weevil world.

Now, in late November, the great mullein flower show is long past. Our early summer mulleins are transformed into brown skeletons, something akin to those Alberto Giacometti sculptures of skinny people that endlessly "stand and wait."

To our surprise, though, this year's great mullein show is far from being over. Many of the seedpods are now enclosing weevils and not mullein seeds. And one thing has a habit of leading to another, especially in our backyard. The downy woodpeckers have cottoned onto these events, and they are having a grand time dining on the weevil babies. This has been a great surprise to us. We are astonished at the strength of the dead mullein stalks, holding up under the weight of the woodpeckers. Even more astonishing is how the woodpeckers are behaving. We are used to these birds hammering at the trunks and branches of trees with great gusto. Instead, the woodpeckers feeding from the mullein play it cool. As gently as a feeding chickadee, the woodpeckers pick at a mullein seedpod and, when in luck, tweak out a tasty weevil snack. I suspect that very few mullein weevils will be emerging next spring.

December

December has a way of scuttling past us, powered by the distractions of national holidays. This is the month for recovery from Thanksgiving celebrations in November and eager anticipation of the holiday season at the end of the year. December brings brightly colored decorations, festive lights, parties with friends and family, and high hopes of a world transformed by a blanket of sparkling snow. With at least one snowstorm a usual part of the December weather mix, these hopes are realistic at some point in the month, although a white Christmas is far from guaranteed. Away from human affairs, the serious business of winter survival in the natural world is all-consuming. For the many creatures that are hiding away, snow cover offers some insulation against plunging air temperatures, whereas active animals, including birds, squirrels, and deer, must negotiate a world that shifts back and forth between freeze and thaw—together with snow, ice, rain, and sun—in their daily task to obtain sufficient food to survive until tomorrow.

Despite the frequently atrocious weather conditions in December, some animals are more than survival machines. Driven by natural selection, they also have sex on the brain. The imperative to find mates is the theme for the opening essay ("Love in a Cold

Climate") and plays a role in the final essay ("Duck Time"). In between, I consider the importance of winter nests for the gray squirrels in our neighborhood ("Squirrel Dreys") and the biology of the coyotes we occasionally hear howling in the night at this time of year ("Coyotes").

Love in a Cold Climate

The temperature was edging up toward freezing as we walked along the upper path overlooking Six Mile Creek. Faint sunlight shone coldly through the thin layer of cloud onto the bare branches of maple and beech trees and the deep green of hemlock. We were making our way to the lookout point over the reservoir, which offers a pretty view of trees and water but rarely has any wildlife of note.

As we approached the lookout, I told myself that the high point of our walk would presumably be the several sightings of flying moths along the way. This was not an especially exciting high point because one would expect to see winter moths on the wing in early December—and also because, let's face it, winter moths are the poster child LBJs of the insect world. They sport a wingspan of an inch (a little less than three centimeters), and they are dull brown-gray. Nevertheless, they are interesting because any insect flying in subfreezing temperatures is decidedly odd.

Various insects can fly in chilly weather. For example, honeybees fly around on sunny days in early spring because they shiver their wing muscles to warm up. Winter moths don't play it that way. Experiments using tiny temperature recorders inserted into the body of winter moths have shown, without exception, that they fly without warming up. The sort-of explanation is that these insects weigh next to nothing and flap their wings slowly. If that explanation doesn't satisfy you, join the club, especially as our winter moths were being no sluggards. Winter moths are unexplained flying objects, the real UFOs of our world.

Our local UFOs may be unexplained, but they are not unidentified. These little moths were most probably males of the Bruce spanworm (*Operophtera bruceata*). They have just one thing on their to-do list: sniff out the females, which don't fly but sit around on trees emitting plumes of an irresistible pheromone. Over the next few weeks, each female will deposit her 150 bright orange eggs in crevices of tree bark. Unless eaten by a chickadee, woodpecker, or other determined bird, the eggs will hatch next spring into ravenously hungry caterpillars that, if in sufficient numbers, can defoliate an entire tree. When fully grown, the caterpillars lower themselves to the ground on a thread of silk, pupating on the forest floor, and, unless eaten by a rodent, shrew, or beetle, will be next winter's cohort of winter moths.

There's an outside chance that our winter moths along Six Mile Creek weren't *O. bruceata* but the close relative *Operophtera brumata*, which is, rather confusingly, known as the winter moth. That would be bad news because *O. brumata* is an invasive from Western Europe and a far more aggressive defoliator than *O. bruceata*. *O. brumata* arrived in Nova Scotia in the 1930s, and it has been spreading into the northeastern US in recent decades. *O. bruceata* and *O. brumata* can't be distinguished by eye, and even the moths themselves have problems because hybrids between the two species have been reported. Entomologists have a solution: the two species can be separated by the detailed morphology of the dissected male genitalia.

I was wrong with my prediction that the winter moths would be the high point of our walk. As we climbed the slope to the lookout, we saw that the reservoir below us was partly iced over, and there was a mass of ducks in the ice-free portion. Many of the birds had a conspicuous white patch on their head, others were brown, and they came in two sizes.

The smaller ducks were seriously small. They had to be buffleheads, ducks that breed in the middle of Canada and migrate through New York State to their winter hangouts farther south. The bufflehead is the smallest duck in North America. Any evolutionary temptation to get bigger is constrained by the preferred nesting site of this species: tree cavities drilled out by woodpeckers. An oversized bufflehead would struggle to get into the cavity to lay and incubate its eggs or feed its young properly. We watched the buffleheads as they dived for food, preened, and generally kept an eye on the world. Then some of the birds took flight, the hazy sunshine illuminating the brilliant white on their bellies, wings, and heads as they streamed up and away.

The other ducks were hooded mergansers, conspicuously larger than the buffleheads. Some of them were behaving just like the buffleheads: resting, preening, and feeding. Then we heard a strange and persistent sound, somewhere between a cat purring, a dog growling, and a frog croaking. It was the ducks—or, more correctly, the drakes. Several of the male mergansers were displaying their prowess to at least one female.

A male hooded merganser on top form is quite something to watch. His crest is raised up and over his head like a hood, displaying the large white patch on each side of his head. It looks almost like a white helmet fringed in black. With his crest in place, the bird rises up, bobs his head forward toward the female, gives it a good shake, and then abruptly jerks his head backward onto the back. Initially, three males were displaying to a single female, but she dived briefly and things got more complicated. A second female was in play, and then more males got in on the act. The diffuse group of romancers drifted along the edge of the ice until they were hidden from our view.

At first, it seems odd to be distracted with affairs of the heart while migrating south. However, hooded mergansers tend not to travel as far as the buffleheads do, and there's no harm in getting an early start with choosing your mate for the next breeding season. It will be a temporary affair, after all. As soon as the female starts to incubate her eggs in April, the male will be off and won't return. This is markedly different from the buffleheads, which, unusually for ducks, maintain a pair bond for years.

We went back to Six Mile Creek a few days later and found that the ducks had gone and the reservoir was entirely empty of birds. All we saw were the winter moths flitting among the trees in the cold.

Squirrel Dreys

The bare winter skeletons of deciduous trees are an annual challenge, requiring us to brush up on tree ID by overall shape, branching pattern, and characteristics of the bark without any leafy clues. It is also an opportunity to discover more about the lives of the gray squirrels that live in our neighborhood.

Our eastern neighbor's box elder tree overhangs our driveway. When the leaves fell, we discovered a brand-new gray squirrel drey. It is high up, as the dreys always are, and perched at the point where a gently upward sloping branch divides into three, creating a small platform. We knew that the squirrels had been busy all summer in the box elder, but we had no idea there was a house construction project in progress.

Drey construction is a summertime job, perhaps extending into September, but its value for the squirrels starts with the rigors of the cold weather in December and beyond. The framework of the leafy nest is a collection of pliable twigs that the squirrel gnaws from the midsummer tree and then weaves into a lattice. Importantly, the squirrel does not remove the green leaves from the twigs. As the season progresses, the leaves turn brown but, isolated from the tree, never receive the hormonal signal to drop. The lattice of leafy twigs might look like a bit of a mess, but it provides protection from rain, snow, and cold. To improve insulation and comfort, the squirrel usually lines the inside of its drey with old grass, moss, and small pieces of shredded bark. In other words, a gray squirrel drey is a big hollow ball, about the size of a watermelon, with a good six inches of living space in the middle. The squirrel builds a single entrance hole into the leafy ball, usually facing down, so that it can race down the branch and tree trunk if disturbed.

Now that temperatures are plummeting—and anticipated to get even colder before winter is done—the box elder residence is a snug lifesaver for at least one of our squirrels. The housebuilder may be joined by one or more other squirrels, most likely close relatives. The warmth of several squirrels huddled together in their well-insulated nest can raise the temperature inside a drey tens of degrees above ambient.

Unsurprisingly, the drey is a high-maintenance dwelling, requiring continual attention to remain intact and waterproofed. This can involve adding more vegetation to fill the cracks and tightening up any parts of the lattice that come loose. The large leaves of the slippery elm at the far end of our backyard are highly prized for drey maintenance, and we routinely see individual squirrels carting these oversized, yellow leaves from the ground up into the trees. If the owner moves away or succumbs to a car or red-tailed hawk, its drey deteriorates rapidly. For example, there's a drey in the large oak tree up the road that looks decidedly lopsided, with twigs and dead leaves hanging down precariously. This is vacant property, I am sure, and the next storm will likely destroy it altogether.

If gray squirrels had homeowner's insurance to worry about, I suspect the builder of our box elder drey would be paying high premiums. Even the most desultory risk assessment would wonder at this ball of twigs and leaves so close to the tip of a branch. Surely, the inhabitants will get motion sickness in high wind, followed by a hefty repair job once the storm has passed. It would have been more sensible to build in a vertical *V* between two stout tree limbs close to the center of the tree. I made a quick check of other dreys in the neighborhood. Of the twenty-four dreys I observed, only three, including our box elder drey, were "out on a limb," as it were. All the other ball-shaped dreys were located in the crotch between the trunk and a sturdy branch or between two substantive branches. We shall see how our box elder drey fares through the coming winter.

This brings me to a further issue about dreys. There are two architectural designs for a drey. One is the big ball design, for a protective residence; the other is the platform design, for lolling about on, especially during the heat of the summer. The latter, often referred to as a rest site, is the squirrel equivalent of budget housing. It's quick to put up, and there are no long-term expectations. Several likely remnants of gray squirrel rest sites were evident on my drey-checking walk.

There is one other kind of gray squirrel residence: the den. All things being equal, I suspect that our imaginary squirrel insurance company would favor a den over a drey. A den is a hole in a tree created, for example, by the drilling of a pileated woodpecker or by the rotting of one side of the trunk after a branch has been torn off in a storm. To my mind, the squirrel specifications for a suitable hole are rather demanding. It must be at least a foot deep and with an entrance about three inches wide. A smaller entryway creates a squeeze for the squirrel; anything much larger and a raccoon can get in. I suspect that holes of the right proportions may often be in short supply. The kingpins of the squirrel world presumably reside in dens, leaving the hoi polloi with no option but to build leafy dreys.

These animals certainly lead complex lives with endless decision-making about their food supply and housing arrangements. The decisions they make may have life-or-death consequences.

Coyotes

Although we live in the city, closely bounded by houses on all sides, it is usually quiet at night. The big exception is the all-night music of the crickets, which is fortissimo through August before gradually diminishing back to silence with the first hard frost or significant snowfall of the year. But one night last week, at about 3 a.m., I heard the unmistakable howling and barking of coyotes. The night air must have been very still, and perhaps, in the cold, the coyotes had strayed closer to town than usual.

A lone animal could not have been responsible. It sounded like a large pack, but a small number of coyotes can make an awful lot of noise. Even the formal name for the coyote, *Canis latrans*, meaning "barking dog," reflects their noisiness. Incidentally, the coyote was named by Thomas Say, the father of entomology who also named the orange-crowned warbler (see October, "A Relocating Crown"), on the same expedition to the Rocky Mountains in 1819–20. Say was a busy man. Back to nocturnal entertainment, the howling pack of coyotes was almost certainly a small family group, probably parents and this year's young, which generally don't disperse till the spring after their birth.

No one has a good thing to say about coyotes. This is obvious from our online neighborhood chat group. Usually, the comments include recommendation requests for local plumbers or babysitters, advertisements for yard sales, notices about missing cats, and updates on the seemingly endless summer roadworks. Every now and again, though, someone posts that there was a coyote in their backyard. It's usually a singleton, almost always described as large, and often called out as "mangy, could be rabid." The description is followed by warnings to keep cats and small dogs indoors in the immediate neighborhood. That's because coyotes have a reputation for enjoying these pets for dinner. (Could it be one of those missing cats from the chat group?) But pets aren't the only thing on the menu. Coyotes also eat rabbits, voles, squirrels, snakes, insects, fruit, and berries. And they won't say no to polishing off any food left on a bird table.

Coyotes are not only bad news. They are also reputed to be clever beasts. There's an online story about how coyotes exterminated a colony of feral cats in Southern California. That might sound like a good thing to some of us, but a local charity was heavily invested in providing feeding stations for the cats. The charity workers didn't realize for ages that their feline friends were no more, presumably courtesy of coyotes, because the food they provided with loving care was always consumed . . . by the wily coyotes.

The coyote's relative, the gray wolf, has been rehabilitated in the minds of many people. Although gray wolves can attack livestock, causing understandable upset among ranchers in the western US, they are widely admired for their complex social lives, hunting skills, and ecological role as a top predator that keeps elk and other deer populations under control. The gray wolf is front and center in all discussions of rewilding the continent.

However, perceptions are different for the coyote. You might imagine that everyone who rails against the white-tailed deer would be glad that coyotes occasionally have deer on their menu. But no! As a neighbor once explained to me, yet another reason for hating the deer in our backyards is that "they attract the coyotes into town." A coyote simply can't win.

One of the problems for our local coyotes is that, in all honesty, they shouldn't be here. That's because the natural range of the coyote was the southwest of the continent, extending into modern Mexico and eastward to the Great Plains. I am rather hesitant about the term *natural range* because it refers to the pre-Columbian range when there were up to 18 million Indigenous people in North America. The ecological footprint of these pre-Columbian humans was far from negligible, and they may have influenced the distribution of coyotes.

Coyotes benefited big time when Europeans pushed west into the continent, persecuting the gray wolf as they went. The problem for coyotes is that they are strongly outcompeted by gray wolves. For example, when gray wolves were reintroduced to Yellowstone National Park in the 1990s, the coyote population declined substantially. Furthermore, studies using radio-collared wolves showed that the wolves chased coyotes away, especially from deer kills, and occasionally killed coyotes. Europeans moving west also changed habitats, replacing dense forest, which coyotes don't like, with coyote-friendly scrubby country and ranchland. There are even stories of fox cubs being sent from the west to fox hunters in the east for release into the areas where they hunted—except that the fox cubs were actually coyote cubs.

And so, the coyote spread eastward, arriving in the Finger Lakes probably in the 1940s. It is a bit more complicated than that. The coyote came not from the west but from the north, Canada. On the way, our coyotes encountered remnant populations of gray wolves. The interactions extended beyond growling and squabbling over dead deer to some matings, such that the coyotes in our area are, genetically speaking, about 26 percent wolf. (For purists, that is 13 percent gray wolf and 13 percent eastern wolf, but people argue about whether the eastern wolf is a subspecies of the gray wolf or a different species.) The genetic admixture is used to explain why the coyote subspecies in the northeast of the continent, including our local patch, is bigger than other coyotes, more prone to staying together in groups and howling loudly, and tends to hunt deer in groups . . . like a wolf.

That's not all. Our local coyotes also have about 10 percent domestic dog in their genomes. In short, coyotes "get about a bit," not only with wolves but also with dogs. Much is made of the so-called coydogs. Although male coyotes are mostly uninterested in domestic dogs, except as food, some male domestic dogs are not as fussy in their affairs of the heart. However, unlike male coyotes, which are superb fathers, dog fathers are strictly looking for one-night stands, leaving the female coyote to raise her offspring alone. This is generally a rather unsuccessful endeavor, but the genomic data tell us that some coydogs survive to adulthood and reproduce.

There is a fair amount of chat about coyotes getting more common in our area. I am uncertain whether this is true. The coyote populations in New York State did increase between the 1940s and 1970s, but, since then, it has been fairly stable at an estimated twenty to thirty thousand individuals (New York State Department of Environmental Conservation n.d.). What is happening is that these clever animals are changing their behavior, increasingly taking advantage of the rich and easy pickings found in human settlements.

There is much discussion about how we can keep coyotes wild. Some say that hunting and trapping is the answer to controlling the numbers and ensuring the animals are scared of humans. I don't know enough to judge whether this is an effective strategy. What does make sense is to yell and gesticulate wildly whenever you see a coyote—to remind them that humans are nasty and should be avoided.

Duck Time

Our local area hosts a total of twenty-four species of ducks, but not all at the same time (Sibley 2003). The distribution maps for these species inform us that just four species are resident species: the mallard and its close relative the American black duck, plus two species of merganser—the hooded merganser and the common merganser. Most of the rest of our two dozen ducks are migrants, passing through our area on their annual trek north to Canada to breed or south to their winter feeding grounds on the coast or on inland ponds and lakes in the southern US and Mexico. For the sake of completeness, I should add that we boast one winter visitor, the common goldeneye, and two summer visitors, the wood duck and the ring-necked duck.

The bottom line of all this detail is that duck spotting should be a low-diversity activity during most months of the year. Luckily for us, nothing could be further from the truth because many species of duck move around during the winter months. Some of these movements are very local, such as finding sheltered places in cold weather. Others are more far-ranging, such as cadging a ride on winds ahead of a cold front. When wind speeds and directions change frequently in unsettled weather, ducks can drop out of the sky to rest and feed on ponds or lakes in unexpected places. This happens again and again. Just as for our visit in early December to the reservoir on Six Mile Creek ("Love in a Cold Climate"), a trip to Cayuga Lake and local ponds can yield a nil return, the same ducks as last time, or a completely different set of species.

We have had a cold snap during the last week of December. It's been nowhere near the record lows for late December, but it's definitely wrap-up-warm-don't-stay-out-for-long weather, hitting 0°F (−20°C) at night and struggling toward the teens during the day (23° to 14°F [−5° to −10°C]), with a gusty westerly wind that gives a windchill that is best not to think about. In these conditions, along with an inch or two of powdery snow, we sprinted around Allan H. Treman State Marine Park at the southern end of Cayuga Lake. The path runs just inland from the lake's edge and behind a line of cattails and other tall reeds. This vegetation gives a bit of protection from the wind and hides us from any birds that are close inshore.

Our bird list for that frigid walk was short but interesting. A continuous layer of ice extended from the shoreline to about twenty yards into the lake. Just beyond the ice, a single pair of buffleheads was cruising back and forth in the chilly water, bobbing from side to side in rhythm with their vigorously paddling legs. Both sported the telltale white patch on the head: hers, an oval patch behind each eye; his, a single large patch extending around the back of the head. Every now and again, one or both lifted its head up and, in an instant, dived down into the water in search of food. At this time of year, buffleheads are said to favor water snails and pondweed.

Apart from the buffleheads, all the ducks were far out on the lake, mostly beyond the resolving power of our binoculars. Our eyes were watering in the bitter wind, but we could make out two sizes of ducks. Curiously, two of the large ones kept on rising up and flapping their mostly white wings (possibly tipped in black, but it was difficult to be sure of that). For the several minutes that we watched, these two individuals repeated this antic, again and again. Our best bet is that they were displaying common mergansers. These birds indulge in what is called courting parties, involving lots of wing flapping, which we could see, together with stretching their bodies out of the water with beaks pressed down to the breast, which we had to imagine. It seems very early to be courting. Apparently, common merganser courting parties generally start in earnest in February or March, although there have been reports of this behavior as early as November. Still, the early date of our displaying birds is not as surprising as the fact that these birds could be thinking amorous thoughts in such bitterly cold conditions.

In the two days following our bufflehead-and-probably-merganser visit, the wind shifted to the south and the temperatures increased to above freezing. We wondered if the change in the weather might have brought any more birds to the lake. When we returned, the ice still formed a continuous layer along the southern shore, but, from a duck's perspective, we were in a different world. There was no sign of our peaceable bufflehead duo. Instead, the place had been invaded by scores of mallard ducks, all quacking, preening, bathing, and generally shuffling about. It sounded like a city park. We wondered where they had been in the cold weather; perhaps farther out on the lake, perhaps taking shelter in the nearby woodland and other protected spots on land.

They weren't all mallards, though. There were several American black ducks among them. The two species are very similar in shape and size, but both sexes of the American black duck are a dull dark brown. Although we can distinguish the mallard and black duck readily, they get confused and hybrids are not uncommon.

We continued scanning the lake. There were lots of small ducks marked with white cheeks extending all the way forward to the beak and a dark cap. Some individuals had a narrow tail of stiff feathers that were stuck up, as if courtesy of hair spray. That made them stiff-tailed ducks, and the ruddy duck is the only plausible stiff-tail in our region. Each bird would be here one moment and then, with the flick of the bill, it would dive down and bob back up like a cork a bit farther along. It is said that the ruddy duck is a supreme diver among ducks. Just before it goes down, it breathes out (yes, breathes out; I would breathe in before diving), tucks its wings and all its feathers close to its body, and then pushes itself forward and down with its legs and outsized feet positioned far back on the body. As is general for diving ducks (but unlike penguins), the ruddy duck doesn't use its wings for swimming.

The ruddy duck is another of the duck species that are described as being found here only on migration from the Great Lakes, where they breed, to the Carolinas and south for the winter. Nevertheless, here were some tens of them making a midwinter visit to our lake during the complex shift in weather conditions.

I find a special pleasure in watching ruddy ducks in their natural habitat. They have a bad reputation in Europe where, after their accidental release during the 1950s, they expanded dramatically and took a liking to mating with the related but rare white-headed duck. As a result, there have been large-scale efforts to cull the ruddy duck and hybrids in the UK and other European countries. At the millennium, there were an estimated six thousand ruddy ducks in the UK, and the number now is between ten and fifteen, at an estimated cost to the taxpayer of 3,000 GBP per bird killed. All sightings of the ruddy duck are required to be reported so any remaining birds can be eliminated. There was a problem, though. Not everyone was happy about this cold-blooded slaughter of ducks, and total extirpation of the ruddy duck may prove to be a difficult task.

Whatever you think of ruddy ducks on the other side of the Atlantic, it is worth a moment to pause and appreciate the fact that there were probably more of them bobbing around the southern end of Cayuga Lake this week than in the entire British Isles. And let's take a second moment to wonder how long they will be there. For sure, they will move on before long, taken by the power of their wings and the winds to . . . who knows where?

Postscript

The natural world is the best show on earth, yet I find it a challenge to construct a simple statement that encapsulates what I value about it. There are two difficulties. The first is that any attempt to distill the essence is ambushed by the detail: the first time each year that I see a nestling bobbing up and down in the osprey nest in the local city park; my greater awareness of goldenrods in meadows and waste places once I knuckle down to identify some of the species; the sparkle of fresh snow in winter sunshine; and the grandeur of a summer thunderstorm.

The second and greater difficulty is that the natural world is not a place of tranquility and ease. Conflict and change are inextricable elements because all organisms struggle to survive and reproduce in a world of variable resources and often unsuitable conditions. Overlaying this biological reality is the outsized impact of one superabundant species—us. The insatiable human demand for space and resources and the unplanned impact of our activities on the global climate are impossible to ignore.

Much is written about the scale of environmental degradation and ongoing mass extinction, and ways to mitigate these anthropogenic changes have been proposed. Most of the solutions are large-scale and few are implemented with any consistency, if at all.

As individuals, it is easy to feel powerless. But despite this seemingly bleak outlook, a small and necessary part of the solution lies in the hands of the individual. Each of us has a responsibility to honor the natural world close to home, an opportunity to revel in the splendor of the familiar.

"Local" is powerful because it is worldwide. Your "local" may be an unknown faraway place for me and vice versa, but, in the tangled bank of this planet, everywhere is local and demanding of respect. I hope that this book on my local natural world helps spark a greater appreciation of your local natural world, wherever that may be.

References

Behler, John L., and F. Wayne King. 1979. *National Audubon Society Field Guide to Reptiles and Amphibians*. New York: Alfred A. Knopf.

Carson, Rachel. 1962. *Silent Spring*. Boston: Houghton Mifflin.

Gillespie, Nina, Suzannah Sherman, Patrick Schröder, Karim Elgendy, and Jon Wallace. n.d. "Cities of the Future." Chatham House. https://www.chathamhouse.org/2022/11/cities-future.

Cornell Laboratory of Ornithology. n.d. "Orange-crowned Warbler Identification." All About Birds. https://www.allaboutbirds.org/guide/Orange-crowned_Warbler/id#.

Darwin, Charles. (1859) 1985. *On the Origin of Species*. There are various editions of this classic text. I use the version with an introduction by J. W. Burrow. London: Penguin Classics.

Darwin, Charles. (1871) 1981. *The Descent of Man, and Selection in Relation to Sex*. I refer to the edition of this text published with an introduction by John Tyler Bonner and Robert M. May. Princeton, NJ: Princeton University Press.

Elhacham, E., L. Ben-Uri, J. Grozovski, Y. M. Bar-On, and R. Milo. 2020. "Global Human-Made Mass Exceeds All Living Biomass." *Nature* 588: 442–44.

Entomological Society of America. n.d. "Spongy Moth Transition Toolkit." https://entsoc.org/publications/common-names/spongy-moth

Franklin, B. 1784. "From Benjamin Franklin to Sarah Bache, 26 January 1784." National Archives. https://founders.archives.gov/documents/Franklin/01-41-02-0327.

Hartig, T., R. Mitchell, S. de Vries, and H. Frumkin. 2014. "Nature and Health." *Annual Review of Public Health* 35: 207–28.

Hopkins, Gerard Manley. 1888. "The Windhover." Poetry Foundation. https://www.poetryfoundation.org/poems/44402/the-windhover.

Jimenez, M. P., N. V. DeVille, E. G. Elliott, J. E. Schiff, G. E. Wilt, J. E. Hart, and P. James. 2021. "Associations between Nature Exposure and Health: A Review of the Evidence." *International Journal of Environmental Research and Public Health* 18: 4790.

Jones, C. G., J. H. Lawton, and M. Shachak. 1994. "Organisms as Ecosystem Engineers." *Oikos* 69: 373–86.

Jordan, K. A. 2022. *The Gayogohó:nǫ? People in the Cayuga Lake Region: A Brief History.* Ithaca, NY: Tompkins County Historical Commission.

Keats, John. 1820. "To Autumn." Poetry Foundation. https://www.poetry foundation.org/poems/44484/to-autumn.

Mabey, Richard. 1996. *Flora Britannica.* London: Sinclair Stevenson.

Marris, Emma. 2011. *Rambunctious Garden.* New York: Bloomsbury.

New York State Department of Environmental Conservation. n.d. "Eastern Coyote." https://dec.ny.gov/nature/animals-fish-plants/eastern-coyote.

Ritchie, Hannah, and Max Roser. 2024. "Half the World's Habitable Land Is Used for Agriculture." Our World in Data. https://ourworldindata.org/global-land-for-agriculture.

Sibley, David Allen. 2003. *The Sibley Field Guide to Birds of Eastern North America.* New York: Alfred A. Knopf.

United Nations. 2023a. "Climate Change 2023 Synthesis Report. *Intergovernmental Panel on Climate Change.* https://www.ipcc.ch/report/ar6/syr/downloads/report/IPCC_AR6_SYR_SPM.pdf (Summary for policymakers) and https://www.ipcc.ch/report/ar6/syr/downloads/report/IPCC_AR6_SYR_FullVolume.pdf (full report).

United Nations. 2023b. "New Analysis of National Climate Plans: Insufficient Progress Made, COP28 Must Set Stage for Immediate Action." https://unfccc.int/news/new-analysis-of-national-climate-plans-insufficient-progress-made-cop28-must-set-stage-for-immediate#main-content (press release).

U.S. Fish and Wildlife Service. 2023. "21 Species Delisted from the Endangered Species Act due to Extinction." https://www.fws.gov/press-release/2023-10/21-species-delisted-endangered-species-act-due-extinction.

Wikipedia. n.d. *Dissosteira carolina.* https://en.wikipedia.org/wiki/Dissosteira_carolina.

Wilson, Karl A. 2014. *Field Guide to the Devonian Fossils of New York.* Ithaca, NY: Paleontological Research Institution Special Publication No. 44.